Sounds and Perception

Sounds and Perception is a collection of original essays on auditory perception and the nature of sounds—an emerging area of interest in the philosophy of mind and perception, and in the metaphysics of sensible qualities. The individual essays discuss a wide range of issues, including the nature of sound, the spatial aspects of auditory experience, hearing silence, musical experience, and the perception of speech; a substantial introduction by the editors serves to contextualize the essays and make connections between them. This collection will serve both as an introduction to the nature of auditory perception and as the definitive resource for coverage of the main questions that constitute the philosophy of sounds and audition. The views are original, and there is substantive engagement among contributors. This collection will stimulate future research in this area.

Matthew Nudds is Professor of Philosophy at the University of Warwick.

Casey O'Callaghan is Associate Professor of Philosophy at Rice University.

Sounds and Perception

New Philosophical Essays

EDITED BY
Matthew Nudds and Casey O'Callaghan

OXFORD
UNIVERSITY PRESS

Great Clarendon Street, Oxford OX2 6DP

Oxford University Press is a department of the University of Oxford.
It furthers the University's objective of excellence in research, scholarship,
and education by publishing worldwide. Oxford is a registered trademark
of Oxford University Press in the UK and in certain other countries

© The several contributors 2009

The moral rights of the authors have been asserted

First published 2009
First published in paperback 2012

All rights reserved. No part of this publication may be reproduced,
stored in a retrieval system, or transmitted, in any form or by any means,
without the prior permission in writing of Oxford University Press,
or as expressly permitted by law, by licence or under terms agreed with the appropriate
reprographics rights organization. Enquiries concerning reproduction
outside the scope of the above should be sent to the Rights Department,
Oxford University Press, at the address above

You must not circulate this work in any other form
and you must impose this same condition on any acquirer

British Library Cataloguing in Publication Data
Data available

Library of Congress Cataloging in Publication Data

Sounds and perception : new philosophical essays / edited by Matthew Nudds and Casey O'Callaghan.
p. cm.
Collection grew out of a conference in 2004.
Includes bibliographical references.
ISBN 978–0–19–928296–8 (hardback : alk. paper) 1. Auditory perception. 2. Sounds. I. Nudds,
Matthew. II. O'Callaghan, Casey.
BF251.S68 2009
128'.3—dc22   2009023183

ISBN 978–0–19–928296–8 (Hbk.)
ISBN 978–0–19–966636–2 (Pbk.)

Contents

Acknowledgements	vi
Contributors	vii

1. Introduction: The Philosophy of Sounds and Auditory Perception 1
   Casey O'Callaghan and Matthew Nudds

2. Sounds and Events 26
   Casey O'Callaghan

3. Sounds as Secondary Objects and Pure Events 50
   Roger Scruton

4. Sounds and Space 69
   Matthew Nudds

5. Some Varieties of Spatial Hearing 97
   Roberto Casati and Jérôme Dokic

6. The Location of a Perceived Sound 111
   Brian O'Shaughnessy

7. Hearing Silence: The Perception and Introspection of Absences 126
   Roy Sorensen

8. The Sound of Music 146
   Andy Hamilton

9. Speech Sounds and the Direct Meeting of Minds 183
   Barry C. Smith

10. The Motor Theory of Speech Perception 211
    Christopher Mole

11. Philosophical Messages in the Medium of Spoken Language 234
    Robert E. Remez and J. D. Trout

Index 265

Acknowledgements

This collection grew out of the Philosophy and Sound Conference at the Institute of Philosophy, School of Advanced Studies, University of London, in conjunction with L'institut Jean Nicod, CNRS, Paris, in 2004. We are grateful to that event's organizers and participants for the opportunity to convene and discuss issues at the core of the philosophy of sounds and auditory perception. Particular thanks are due to Tim Crane and to Barry C. Smith.

During the preparation of this volume, Chapter 7, 'Hearing Silence: The Perception and Introspection of Absences', appeared as chapter 14 of Roy Sorensen's *Seeing Dark Things* (Oxford University Press, 2008).

Innumerable thanks to Peter Momtchiloff.

Contributors

ROBERTO CASATI, Senior Researcher, Centre National de la Recherche Scientifique, France

JÉRÔME DOKIC, Professor, École des Hautes Études en Sciences Sociales; Member, Institut Jean-Nicod (CNRS, EHESS, ENS)

ANDY HAMILTON, Senior Lecturer in Philosophy, Durham University, and Adjunct Lecturer, University of Western Australia

CHRISTOPHER MOLE, Lecturer in Philosophy, University of British Columbia

MATTHEW NUDDS, Senior Lecturer in Philosophy, The University of Edinburgh

CASEY O'CALLAGHAN, Assistant Professor of Philosophy, Rice University

BRIAN O'SHAUGHNESSY, Emeritus Reader in Philosophy, King's College London

ROBERT E. REMEZ, Professor of Psychology, Columbia University

ROGER SCRUTON, Research Professor, Institute for the Psychological Sciences

BARRY C. SMITH, Professor of Philosophy and Director of Institute of Philosophy, University of London

ROY SORENSEN, Professor of Philosophy, Washington University in St. Louis

J. D. TROUT, Professor of Philosophy, Loyola University Chicago, and Adjunct Professor, Parmly Sensory Sciences Institute

1

Introduction: The Philosophy of Sounds and Auditory Perception

CASEY O'CALLAGHAN AND MATTHEW NUDDS

1. Sounds and Perception

'Humans are visual creatures', it is common to observe. Our reliance upon vision is apparent in the way we navigate and react to our surroundings. We fumble in the dark and instinctively turn to look at the sources of sounds. Visual information also occupies a privileged epistemic role, and our language frequently reflects a tight coupling of seeing with knowing. We evaluate *views*, have *insights*, and *see* what is at issue. Perhaps most telling is the greater fear many admit at the prospect of losing sight over any other sense.

Not surprisingly, philosophers investigating the nature of perception and perceptual experience have considered vision nearly exclusively. Philosophical discussions of sensible and secondary qualities have focused upon color and color experience, while debates about perceptual content primarily concern the content of visual experiences.

Until remarkably recently, something similar was true of empirical researchers who aimed to unearth the processes, mechanisms, and principles that explain how we become acquainted with our environments. Driven by the goal of computer vision, vision scientists were among the first to shed sensory psychology's early preoccupation with psychophysics and the measurement of sensations. Empirical work on perceiving and attending to visual objects has since advanced to the point that Brian Scholl (2001: 2) has described it as 'a type of "case study" in cognitive science'. Vision is better understood than any other sense modality.

But humans are not solely visual creatures. Exclusive attention to vision distorts the degree to which we rely on each of the senses to cope with information-rich surroundings. Recently, interest has grown rapidly in understanding the other sense modalities and sensible features that figure in our capacity to negotiate and understand our environments. Spurred in part by a growing body of rich empirical research, philosophers increasingly have turned attention to tactile, proprioceptive, and kinaesthetic perception; smell and olfactory experience; and aspects of the philosophy of taste (see, e.g., O'Shaughnessy 1989; Martin 1992; Scott 2001; Gallagher 2005; Lycan 2000; Batty 2007; B. Smith 2007). The 'other' sense modalities present challenging new puzzles for the empirical and philosophical study of perception.

No topic in extra-visual philosophy of perception has generated as much attention in recent years as that of sounds and audition. While Strawson (1959) set an early example in *Individuals* by exploring the conceptual consequences of a purely auditory experience, and Evans (1980) responded with a revealing discussion of the requirements on objective experience, the past decade has seen a flurry of work on the nature of sounds and the content of auditory experience. Current research on the perception of speech sounds and spoken language, the experience of music, auditory-visual cross-modal illusions, and the nature of 'auditory objects' promises to impact and advance the philosophy of perception.

More important, however, it signals a departure from the tradition of relying upon vision as the representative paradigm for theorizing about perception, its objects, and its content. While the implicit assumption has been that accounts of visual perception and visual experience generalize to the other senses, nothing guarantees that what is true of seeing holds of touching, tasting, or hearing. Intuitions about critical issues or particular cases might differ in the context of different modalities. While it might seem obvious in the case of vision that perceptual experience is *transparent*, or that space is required for *objectivity*, gustatory and olfactory experiences might tell otherwise (see, e.g., Lycan 2000; A. D. Smith 2002).

Furthermore, resolving certain issues might require examining modalities other than vision. For instance, the debate whether the phenomenological characteristics of experiences are a subset of their representational properties turns on whether visual and non-visual experiences that share representational properties share phenomenological character. Resolving this question depends upon whether it is plausible that all non-visual experiences have representational content, whether visual and non-visual experiences can share representational content, and how best to characterize the phenomenology of non-visual experiences. Given the present state of debate, whether intrinsic properties of experiences constitutively contribute to their phenomenology might only be apparent upon considering experiences in other modalities and phenomenological differences among modalities.

Even if one's sole concern is vision, examining the other modalities enriches one's understanding of what it is to perceive visually and of how we ought to characterize the phenomenology and content of visual experience. Debates about vision and visual experience are informed by attention to other sense modalities.

Some cases even indicate that one cannot give a complete account of perceiving in any single modality without appreciating phenomena that involve other modalities and without addressing the relationships among the senses. For instance, given an important class of inter-modal effects and cross-modal recalibrations and illusions, the content of vision might in certain respects depend either upon the content of experiences that take place in other modalities or upon amodal content that cannot be characterized exhaustively in purely visual terms. In either case, information associated with another modality impacts experience in vision and helps to determine its content. Whether the relationship between extra-visual information and visual experience is constitutive, merely causal, or entirely accidental, a complete accounting that explains these visual processes and experiences requires understanding of the other senses and the relationships among modalities.

In addition to helping advance familiar debates in the philosophy of perception, the case of sounds and audition reveals new puzzles. One example is whether and, if so, how we hear anything but sounds. For instance, when a door slams, I hear its sound. But I also seem to hear the slamming of the door. The slamming is what motivates me to react. So, while I hear the sound of a door slamming, is it also fair to say that I hear the door itself? If so, how do things other than sounds enter into the contents of auditory experiences, and what is it to auditorily represent a door? Alternatively, are the sources of sounds perceived only indirectly thanks to one's awareness of sounds?

Another example involves the nature of sounds themselves. Traditionally, sounds have been grouped with the colors, tastes, and smells among secondary or sensible qualities. Recently, however, a number of philosophers have argued that sounds are not qualities or properties at all, but instead are events. On this account, sounds are more analogous to visual objects than visible features, in that sounds are the bearers of audible features. This raises a number of questions. If sounds are events, what is it to experience an event in a way that does not depend upon experiencing its participants? Do we experience a source to generate or cause a sound?

The philosophy of sounds and audition also opens new fronts in the philosophy of perception. Considering sounds and hearing forces philosophers to confront the cases of music and spoken language. Listening to music and perceiving speech provide fascinating examples of hearing's richness and complexity. The possibility of an emotionally engaging temporal art of sounds and the existence of a fluid and flexible communicative medium comprising sounds illustrate the extent to which audition is a significant and central perceptual domain that should not be ignored by the philosophy of mind and perception.

This collection comprises original essays that address the central questions and issues that define the emerging philosophy of sounds and auditory perception. This work focuses upon two sets of interrelated concerns.

The first is a constellation of debates concerning the ontology of sounds. What kinds of things are sounds, and what properties do sounds have? For instance, are sounds secondary qualities, physical properties, waves, or some type of event?

The second is a set of questions about the contents of auditory experiences. How are sounds experienced to be? What sorts of things and properties are experienced in auditory perception? For example, in what sense is auditory experience spatial; do we hear sources in addition to sounds; what is distinctive about musical listening; and what do we hear when we hear speech?

This introductory chapter has three aims. It presents a survey to provide context for the issues discussed in the chapters that follow. It summarizes the main debates and arguments at stake in this volume. And it suggests promising areas for further work, including unsettled questions and topics that remain unaddressed.

2. The Ontology of Sounds

A theory of sounds should identify the ontological kinds to which sounds belong, and it should say what sorts of properties sounds possess. Debates about the nature of sounds have focused upon such questions as whether sounds are mind-dependent or mind-independent, whether they are individuals or properties, and whether they are object-like or event-like. Also, there has been considerable debate about just where sounds are located.

2.1 What Kind of Thing is a Sound?

Sounds are among the things we hear. Auditory experience is directed upon sounds. Sounds, therefore, are *intentional objects* of audition (see Crane 2009). Since it is plausible that sounds are perceived only through the sense of hearing, sounds commonly are counted as *proper* sensibles of audition. Furthermore, it is plausible to say that whenever you hear something, and whatever you hear, you hear a sound. It is doubtful you could hear something without hearing a sound. Arguably, this is because whatever you hear—such as a collision or a trumpet—you hear it by or in virtue of hearing its sound. Sorensen (Chapter 7), however, disagrees. He argues that we hear silence, which does not involve hearing a sound. Traditionally, nevertheless, sounds are counted among the *immediate* objects of audition.

Given their status as immediate and proper objects of audition, it is not surprising that the nature of sounds has been tied to our experience of sounds. Since at least the early modern era, the predominant view has been that sounds are secondary or sensory qualities. Locke, for one, grouped the sounds with the colors, tastes, and smells as dispositions whose characterization tied them essentially to the experiences of subjects. In the 20th century, some theorists held that sounds are subjective and private and that they mediate auditory perceptual access to the world (e.g., Maclachlan 1989).

Sounds, however, need not be counted as private and subjective given their status as immediate objects of audition if we reject that perception enlists subjectively accessible intermediaries, as do contemporary representationalists along with direct realists and disjunctivists (see, e.g., Tye 2000; Noë 2004; Martin 2006). Sounds then might be experientially or subjectively immediate, which allows either that perception involves no mediators (including representations) at all, or that it requires no experientially accessible but subjective or private mediators.

Sounds might still be grouped with other *perceptible qualities* or *properties*, such as colors, smells, and tastes. For instance, Pasnau (1999) argues that sounds are properties that either are identical with or supervene upon vibrations of things such as bells. On this account, sounds are properties attributed to things commonly taken to be the sources of sounds.

Some recent philosophers have argued that sounds are not properties or qualities, but instead are *individuals* or *particulars*. Rather than *qualifying* or *being properties attributed to* things, sounds are individuals that bear sensible features such as pitch, timbre, and loudness. Sounds on this view are not mere dimensions of similarity.

O'Callaghan (Chapter 2; see also 2007), for instance, claims that property theories do not capture the individuation and identity conditions for sounds. O'Callaghan claims that sounds persist through time and survive changes in ways that sensible qualities and features do not. This raises the question whether sounds are object-like individuals or event-like individuals. O'Callaghan argues that sounds do not simply persist, but have *durations* and commonly are individuated in terms of the features they exhibit *over time*. For example, the sound of a police siren comprises a certain pattern of changes in audible features over time. The sound of the spoken word 'siren' differs from that of 'silent' in that the two involve different patterns of change through time. So, many sounds are individuated in terms of patterns of features over time. This, and the difficulty of imagining an instantaneous sound, suggests sounds are essentially temporal.

Impressed by the temporal natures of sounds, several philosophers have argued that sounds are *events* of a certain kind. Casati and Dokic (Chapter 5; see also 1994, 2005) identify sounds not with the property of vibrating, but with the *event* of an object's vibrating. O'Callaghan identifies sounds with a closely related but different event. O'Callaghan argues that the presence of a medium is a necessary condition not just upon the perceptibility but upon the existence of a sound, and proposes that sounds are events in which vibrating objects or interacting bodies actively disturb a surrounding medium. This account differs from Casati and Dokic's in three ways. First, sounds are not identical with vibrations. Either they are causal byproducts of vibrations, or they are vibrations only under certain conditions. Second, sounds may result from events such as collisions or strikings in which multiple objects interact. Finally, sounds require a medium and thus cannot exist in a vacuum.

Scruton (Chapter 3; see also 1997) offers a very different kind of event theory of sounds. Scruton rejects the physicalism of Casati and Dokic and O'Callaghan, and argues that sounds are what he calls *secondary objects* and *pure events*. First, on analogy with secondary qualities, sounds, like rainbows and smells, are *secondary* objects of perception. Secondary objects, unlike secondary qualities, are independent particulars or individuals rather than properties or qualities. But, like secondary qualities, they are not identifiable with any physical features or objects. The features of such individuals include just their ways of appearing. Secondary objects are objective, though simple and irreducible. Scruton also claims sounds are *pure* events that do not happen *to* anything and that cannot be reduced to changes to other reidentifiable particulars. Sounds thus lack a constitutive ontological connection with the vibrations or activities of objects we ordinarily count as sound sources. Appreciating the independence of sounds from sources, according to Scruton, is critical to understanding distinctively musical experiences: hearing music *requires* the ability to experience sounds as independent from their physical causes (see Section 4.2 below).

Perhaps surprisingly, none of these accounts constitutively ties sounds to longitudinal pressure waves that pass through an elastic medium such as air or water or metal. Such waves propagate from their sources outward towards observers, have frequency and amplitude, and cause auditory experiences.

According to common sense tutored by science, sounds just are traveling waves.

Several authors in this collection, including Nudds (Chapter 4), O'Shaughnessy (Chapter 6), Sorensen (Chapter 7), and Smith (Chapter 9), endorse theories inspired by the common scientific account. Sorensen, for instance, says, 'Since I identify sound with acoustic waves, I think silence is the absence of acoustic waves' (p. 140). Nudds argues that even though sounds are not identical with waves, they are dependent upon waves. More carefully, he argues that sounds are *instantiated* by waves. According to Nudds, sounds, such as those of words or symphonies, can be instantiated on different occasions and by different waves and frequency patterns. Nonetheless, we may perceptually identify a sound as the very same sound whenever it is instantiated. Nudds thus claims that sounds should be understood either as particularized types or as abstract particulars that are instantiated by the waves. The virtue of this account is that sounds themselves are repeatables, but they are not features of waves, a medium, or objects. This view preserves the intuition that we can make or hear the same sound on multiple occasions while rejecting the claim that sounds simply qualify their sources.

2.2 The Locations of Sounds

One main disagreement between the wave-based accounts of sound such as those of Nudds, Sorensen, and O'Shaughnessy (see also Hamilton, Chapter 8) and source-based accounts such as those of Pasnau, Casati and Dokic, and O'Callaghan (see also Matthen forthcoming) concerns the locations of sounds. The former locate sounds in the medium and imply that sounds propagate and thus occupy different locations over time, or travel. The latter hold that sounds are located at or near their sources and do not travel through the medium—sounds travel only if their sources do.

Debate surrounding this issue draws attention to a substantive constraint on theorizing about sounds and their natures. How we experience sounds to be serves as a prima-facie basis for any account of sounds. This is because, in the first instance, our access to sounds is through auditory experience, and our conceptions of sounds are grounded in experience. An account of sounds should be an account of things it is plausible to identify with sounds as we experience them to be. How our experiences of sounds present them to be thus constrains what account it is plausible to give of the nature of sounds. One way to formulate this experiential constraint on theorizing about sounds appeals to veridicality. An account of sounds should entail that auditory experiences of sounds are for the most part veridical; all else equal, it should not imply that experiences of sounds involve wholesale illusions. So, we might hold that for any feature sounds are experienced to have, it at least is possible for experience to be veridical in that respect. A weaker version holds that, even if the experience of a sound could not be veridical in all respects, sounds should have at least most of the features we experience them to have. This means that, all else equal, for some feature we experience sounds to have, we should prefer an account that does not ascribe illusion with respect to that feature. We can put the constraint as a slogan: avoid attributing unnecessary illusions.

Distal sound theorists commonly argue that sounds seem in auditory experience to be located at or near their sources. Sounds, they claim, do not seem travel from the source towards your ears, do not under ordinary conditions seem to pervade the medium (perhaps they do under special circumstances, such as in a loud nightclub), and do not seem to be nearby or at the ears. Instead, they claim that sounds auditorily seem to be where the things and events that generate them are located. If we do experience sounds to be distally located, and if sounds are roughly where they seem to be, then sounds do not travel through the medium as wave accounts imply. Distal theorists charge that unless we systematically misperceive the locations of sounds, sounds do not travel through the medium as do pressure waves (Pasnau 1999; O'Callaghan, Chapter 2). In that case, the veridicality constraint means that we should favor the distal view. Hamilton disagrees, and argues that we hear only where the traveling sounds have come from, rather than where they are. A related response is that we hear, veridically, only a subset of the locations of sounds.

The distal theories support an account according to which auditory perception is in important respects analogous to vision. In particular, sounds located at a distance are perceived thanks to a medium (pressure waves) that bears information about them. Sound waves on this account are like the light that conveys information about distal objects and stimulates vision. The physical waves are not the sounds, and the sounds do not travel with the waves, but the waves mediate between sounds and hearers.

On the other hand, some authors maintain that auditory perception differs in this respect from vision. Suppose that in audition we experience a sound that is proximal when we experience it, and that, in virtue of experiencing the sound, we perceive something that is distal. On this account, the sounds heard are located near their perceivers, but they provide information about distal things and events beyond the world of sounds. Such a proximal theory of perceived sounds preserves the metaphysical dependence of sounds upon the sound waves that stimulate hearing. In effect, it locates the sounds we hear (at the time we hear them) at a different stage in the causal chain that leads from source to subject. That causal chain begins with the activities of things in the environment, leads to wave-like motion in a medium, continues with stimulation of the auditory sense organs, and culminates in auditory experiences. Distal theories locate the sounds we hear at an earlier stage in the causal sequence than do proximal theories.

Since proximal theorists do not wish to say that auditory experiences involve a systematic spatial illusion, they must reject the distal theorists' phenomenological claim that sounds seem in audition to be located at a distance in some direction. Proximal theorists and distal theorists therefore disagree about how best to describe the spatial aspects of auditory experience. Resolving the issue therefore requires a closer examination of spatial audition.

3. Spatial Audition

It would be difficult to deny that hearing conveys spatial information. On the basis of audition, you learn that the barking dog is behind you or that the door to your left has closed. But we can explain in different ways how you learn this. Distal theorists argue that you hear sounds to be located at some distance in a given direction and thereby come to learn about, and perhaps even to hear, the locations of their sources. At the other extreme, subjects might merely *infer* or *work out* information about space and locations from entirely aspatial auditory experiences (O'Shaughnessy, Chapter 6; see also 1957 and Malpas 1965). Smith says: 'Sounds, in general, are hard to place in the spatial world and auditory perception gives us no clues as to where they might occur' (p. 202). The disagreement concerns whether or not audition itself involves perceptual awareness of spatial characteristics, and to what it attributes those spatial characteristics.

Skepticism about spatial audition has been widespread at least since Strawson's (1959) famous claim that a purely auditory experience—in contrast to a purely visual or purely tactile-kinaesthetic experience—would be entirely non-spatial. Strawson claims that a world of sounds would be a no-space world because sounds are not intrinsically spatial. According to Strawson, spatial concepts have no intrinsically auditory significance, and audition's spatial capabilities depend upon its inheriting spatial content from other modalities.

While Strawson's arguments are subject to different interpretations and have been challenged (see Nudds 2001; O'Callaghan forthcoming; Casati and Dokic), they suggest an alternative way to understand how audition grounds spatial beliefs. First of all, not all contemporary proximal theorists wish to deny the vast body of research showing that for perceptually normal subjects with vision, touch, etc., under ordinary circumstances with multimodal stimulation, hearing itself is spatial. Under such conditions, auditory experience might have spatial content or represent spatial features (see Blauert 1997; Nudds, Chapter 4; Casati and Dokic, Chapter 5), whether this depends upon other modalities or not. Nonetheless, one might claim that we do not experience *sounds* as having spatial features. Nudds, for instance, argues that sound *sources*, rather than sounds themselves, are auditorily experienced as distally located. This accommodates the empirical evidence about auditory localization without accepting that sounds themselves are experienced to be located. On his account, information embodied in sound waves about the locations of sound sources is used to determine and auditorily represent the locations of sound sources without representing sounds as distally located. Such an account might go on to claim that sounds seem to be located at or near the ears, that they seem nearby but to have *come from* some direction, or that they seem to lack spatial features entirely. Therefore, although audition has spatial content, it need not attribute spatial properties, such as distal location, to sounds. Sounds might seem nearby or nowhere, while sound sources seem located at a distance.

Distal theories maintain that information about the locations of sound sources is provided by the audible locations of sounds *at* their sources. In contrast, some proximal theories that attribute spatial content to auditory experiences hold that audition attributes spatial properties to sound sources. Both proximal and distal accounts thus may hold that auditory experiences have spatial content, or that spatial properties are represented in audition. But they may disagree about that to which spatial properties are attributed.

Two things are worth noting. First, the kind of proximal account just described owes an explanation for how audition could represent sound sources as having spatial characteristics without representing sounds as located or as having spatial features. How could sound sources auditorily seem located distally if sounds do not?

Second, in considering where sounds are located, we need to consider where sounds are experienced to be located. This, in turn, leads to a discussion of spatial audition. The facts about spatial audition, including the auditory experience of spatial features of an environment, however, appear to be compatible with the view that we hear sound sources, rather than sounds, as located. Evaluating this alternative to the claim that sounds are distally located thus forces us to consider what audible attributes ordinary objects and events that generate sounds, such as bells and collisions, possess. The proposed account requires that, in addition to the sounds, we are capable of auditorily perceiving the sources of sounds. While distal theories may allow for the auditory perception of sound sources, their account of spatial audition does not obligate them to do so.

So, there are different options if auditory spatial beliefs about the environment are grounded in spatial audition. First, we auditorily experience distally located sounds, perhaps along with their sources. Second, we hear sounds locally or aspatially, but thereby experience distally located sound sources. Alternatively, one could deny (implausibly, in our view) that auditory experience itself has spatial content.

It is doubtful whether introspection of auditory experience alone could decide among these possibilities (see, especially, Schwitzgebel 2008 for doubts about phenomenological introspection; see also Remez and Trout, Chapter 11, discussed below in Section 4.3). Audition provides lots of useful information about things and happenings that generate sounds. Indeed, one way to individuate sounds appeals to their causal sources. Experiences of sounds thus are closely associated with perceptual information about their environmental sources. Reflecting upon the phenomenology of spatial experience alone may not be decisive without some independent way to determine where we experience sounds to be and whether we auditorily experience sound sources.

The dispute over the locations of sounds thus turns on a family of questions about the content of auditory experience. In addition to sounds, do we hear their sources? Which properties—in particular, which spatial properties—does audition attribute to each? Progress on these issues requires a more detailed study of the content of auditory experience.

4. The Content of Auditory Experience

4.1 Sounds and Sources

Accounts of the content of auditory experience can be sorted into three classes. First, austere views hold that we immediately hear only sounds and their attributes, such as pitch, timbre, loudness, duration, and location. Second, more permissive accounts hold that we might hear both sounds and their sources. According to such accounts, we might hear the sound and hear the bell or its striking. In that case, we also might auditorily experience sounds in some sense to *belong* to their sources. For instance, sounds might be heard as properties or as parts of their sources. Alternatively, sounds might be heard to be distinct from their sources, in which case we also might hear the relations between sounds and sources. Third, an account could maintain that we hear even things beyond sounds and their sources, such as how things stand in the environment. For instance, in hearing the sound of footsteps I might also hear the enclosed space in which they are being taken.

Deciding among these options poses a methodological challenge. We might appeal to what we say we can hear, or to what we can learn on the basis of hearing. Typically, we say we hear the bird singing as well as the sound it makes. We report learning about the locations of sound sources such as cars or collisions on the basis of hearing. But, with vision, we can say we see that the mail carrier has come on the basis of seeing the pile of mail without being committed to claiming that visual experience represents that the mail carrier has come. So, perhaps we can say that we hear that the bird is singing on the basis of hearing the sound, without saying that auditory experience represents the bird. In general, we need to distinguish what is part of the content of experience from what we learn or judge based upon experience. Though we learn about the sources of sounds on the basis of hearing, appealing to what we can normally come to know on that basis is not an infallible guide to the content of auditory experience.

While we might appeal to the phenomenology of auditory experience to support one or another account of its content, we turned to considering the content of auditory experience in part to avoid relying entirely upon phenomenological introspection. Nonetheless, there are considerations that support thinking that awareness as of sources is an important part of the content of auditory experience. It would be difficult otherwise to explain why we so persistently form beliefs about the sources of sounds on the basis of audition without inference or further assumptions, and it would be difficult to account for the fact that we act on the basis of auditory experience as if we heard sound sources. Reflexively turning to look for the source of a sound or ducking when you hear something coming from behind would make little sense unless you were aware of sound sources. Furthermore, we could make a strong case that your auditory experience as of the sound of a bell would not be veridical if you opened your eyes to see a loudspeaker or a duck.

We might appeal to a general metaphysical view about the nature of perceptual experience, such as a sense datum view (which perhaps favors an austere account) to decide the issue. However, the goals of theorizing about audition and sounds include testing such accounts and learning if they generalize. Furthermore, most contemporary accounts of perception, such as direct realism or intentionalism, are compatible with each of the options.

Another alternative is to appeal to the function of auditory perception and to the kinds of psychological explanations into which auditory content enters.

Though the bulk of laboratory work on audition has used artificial tones in artificial situations, a growing body of work on ecological psychoacoustics appears to provide support for the claim that how auditory processes deal with acoustic information depends in important ways upon natural constraints that amount to assumptions concerning the physical world and properties of sound sources (Neuhoff 2004). For instance, features of *sources*, such as material and size, which determine how they vibrate and disturb the medium, explain dimensions and degrees of auditory similarity and difference that acoustic characteristics alone cannot (see, e.g., Handel 1993; McAdams and Bigand 1993; Bregman 1990). For instance, explaining timbre perception probably requires appeal to features of sound sources (see Handel 1995). This supports a compelling conception of the role of audition as furnishing awareness of the things and happenings in our environments that make sounds.

Perhaps unsurprisingly, then, a prominent theme throughout this collection is that awareness of sound sources is an important aspect of auditory experience. Several contributors here reject the austere claim that we immediately hear only sounds and so must infer or judge what produced them (e.g., Nudds, Chapter 4; Hamilton, Chapter 8; Smith, Chapter 9).

Those who endorse that sources are part of the content of audition do not *just* hold that in addition to hearing sounds, we hear the things that in fact are their sources. Rather, they generally hold some view about the relationship we hear sounds to bear to their sources. Co-location is one such relation (as are other spatial relations). Another possibility is that sounds are heard to be properties of or to qualify their sources. This option is unavailable to those who reject property views of sounds for reasons such as those outlined above. Another possibility is that sounds are heard to be mereological parts of their sources (O'Callaghan 2008). On such a view, sounds might be heard to be parts of events that involve ordinary objects such as bells and whistles. For instance, two cars are involved in a collision, and part of that event is a sound. Hearing a collision by hearing its sound might be akin to seeing a cube by seeing its facing surface (cf. Scruton). A final possibility, according to which sounds are heard to be caused or produced by their sources, perhaps fits best with ordinary thinking about sounds. This requires that we are able to perceive causal relations. It also requires experiencing sources as independent from their sounds, so it remains to explain how we are perceptually aware of sound sources as such.

If audition does involve awareness of sound sources, then audition differs in an important respect from vision. One's auditory awareness of sound sources intuitively is not as *direct* as when one sees those same sources. Thus, even accounts on which we hear distally located sounds, if they also allow that we hear sound sources, might imply that audition involves a form of awareness of sources that is less direct than visual awareness of objects. This invites a new discussion of the ways in which perception may be direct or indirect that extends beyond the visual case.

It is noteworthy that so many have found it compelling that auditory awareness does not stop with sounds. This contrasts with vision, where fewer have been inclined to say that we see what is causally upstream from the objects, colors, and shapes we visually experience. In the visual case, accounts of 'seeing in' and 'seeing as' and 'metaphorical seeing' commonly are invoked. In contrast, hearing a bell or a bird that makes a sound requires no similar act of imagination.

Considering whether sounds or sound sources are auditorily experienced as located led us to consider whether audition involves awareness of sound sources. While it is not obvious that we auditorily experience sound sources, there are some reasons for thinking that we do. Obstacles remain. What relationship are sounds experienced as bearing to their sources? What features can sound sources be auditorily experienced as having? Why acknowledge indirectness in audition if not in vision? This debate cannot yet settle the question about the locations of sounds. However, it does impact how we should characterize auditory experience, and it raises more questions than it answers about auditory content.

4.2 Music

An account of human auditory perception should accommodate music. Since speech raises special questions that we will address in turn, consider *pure* or non-vocal music. The possibility of an art of non-vocal sounds raises special questions about the nature of musical listening. Does hearing music require a distinctive act of listening? What is aesthetically significant about listening to pure music? This depends upon what is aesthetically significant about music. Since, presumably, we are capable of hearing at least some aesthetically relevant features of music, it also depends on the content of our auditory experience of music. Because it is prima-facie plausible to think that the aesthetic significance of pure music depends only upon sounds in abstraction from the environments or circumstances of their production, however, the case of music contrasts with the case of ordinary audition. This contrast may illuminate non-musical auditory experience. Reflecting on musical listening may, therefore, provide evidence to help resolve the questions about auditory content addressed above.

Is listening to music just a variety of ordinary auditory experience, or is it special? For instance, does musical listening require unique or special capacities or skills? On one hand, music involves sounds and sound sequences, arrangements, or structures. So, ordinary auditory capacities are needed for hearing music. If music is nothing more than sounds, such capacities should suffice. However, it *is* plausible that one could perceive auditorily without hearing music as such. Animals, for instance, might hear sounds without experiencing music. Musical experience might involve more than just hearing sequences of sounds. But the difference could just be a matter of how one *responds* to one's auditory experience. For instance, music often incites emotions, imaginations, or associations that are triggered by hearing patterns of sounds. Nevertheless, such responses are responses either to a distinctive variety of auditory experience or to particular aspects of one's auditory experience. What are the features of auditory perceptual experience when listening to music that make possible the distinctive experience of music?

Roger Scruton (Chapter 3; see also 1997) argues that musical listening requires hearing in a way that abstracts from one's interest in the environmental sources of sounds. According to Scruton's *acousmatic thesis*, humans' capacity to appreciate music depends upon the unique ability to auditorily experience sounds as detached entirely from their physical causes, or as divorced from the worldly sources of their production. The aesthetic characteristics of music, according to Scruton, are independent from such facts as that individual sounds are produced by an oboe, or a particularly rare oboe, or that a passage requires a high level of skill to perform, or that a performance is live rather than recorded. What matters are the sounds. (Recall, for Scruton, sounds are secondary objects and pure events that are independent of their sources.) This account of musical listening requires that in some sense it is possible to have auditory experiences whose contents include sounds but not sound sources. On a strong reading, listening to appreciate the aesthetic character of music requires auditorily experiencing sounds, without experiencing their sources. That would seem to require an austere, sound-only account of auditory content. One alternative is to deny that musical listening requires austere auditory content and to hold, instead, that musical listening is a matter of *attending* to that which is (independently motivated to be) aesthetically relevant, to wit, the sounds. This modification requires only the capacity to experience and attend to sounds *as* independent from their sources, rather than the capacity to experience sounds without experiencing their sources.

Andy Hamilton (Chapter 8; see also 2007) resists Scruton's acousmatic thesis that musical experience involves awareness of sounds that divorces them from their sources, and argues that attending to sounds as part of the world in which they are produced is an aesthetically relevant aspect of musical experience. Hamilton offers a twofold account on which acousmatic and non-acousmatic listening both provide valuable musical experiences. Hamilton suggests that features that outstrip sounds, such as the skill of a performer, or the fact that sounds are produced by a performance rather than by a recording, can be aesthetically relevant. Since Hamilton holds that many such features, in addition to sounds themselves, can figure in auditory experience, he argues that auditorily experiencing music involves non-acousmatic experiences. Hamilton thus holds that there is a sense in which we can hear the production of sounds through hearing alone. Sources therefore must enter the contents of auditory experience on this view of musical experience.

But Hamilton also holds that the experience of music is not purely auditory. First, there are aesthetically relevant features of music that we experience through senses other than hearing—including sounds! 'We feel as well as hear sounds' (p. 166), and we see as well as hear the virtuosity of a performance. Such extra-auditory experiences must be non-acousmatic. Moreover, Hamilton doubts whether even acousmatic experience must be purely auditory and thus unimodal. Given multimodal influences that shape perception, listening to sounds in a way that abstracts entirely from their sources, and from other senses, may prove impossible. In that case, multimodal or amodal aspects of perceptual experience may unavoidably infect auditory experience. In that case, even 'purely' auditory experiences of sounds might have non-acousmatic features.

Scruton would simply resist that non-acousmatic aspects of auditory experience are relevant, and he might reject that the other senses matter to our appreciation of music. But, *if* auditory experiences of sounds *unavoidably* have non-acousmatic features, then the acousmatic thesis as stated requires revision. Scruton might comfortably speak of the *aspects* of auditory experience, or of the *features* of sounds, that are aesthetically relevant. This, however, is compatible with rejecting that a special mode of musical listening exists. If musical listening is a unique variety of auditory experience, perhaps it involves a distinctive way of aligning auditory attention. In that case, the skillful act of musical listening could be like an abstract or formal, non-representational mode of looking at paintings or pictures, a way of looking that involves an appreciation of the arrangements of colors and pigments rather than of what is depicted. Arguing that attending to formal features of sounds is the only aesthetically significant way of listening to music raises questions similar to those raised by the corresponding claim about looking.

4.3 *Speech*

Just as humans, perhaps uniquely, are in a position to hear sounds as music, we also may be unique in hearing sounds as speech. Speech, like music, raises questions about the contents of auditory perceptual experiences. In particular, to what extent do the experiences of hearing speech and of hearing ordinary environmental sounds share auditory perceptual content? However, the case of speech also introduces complexities that force us to reconsider whether sounds are among the objects of speech perception. Moreover, some researchers even maintain that speech perception is a unique perceptual modality. Thus, the philosophical issues about speech perception concern different versions of the question: Is speech special?

Hearing and appreciatively listening to music involves focusing on acoustical properties of sounds. Perceiving spoken language, however, requires not just hearing sounds, but also grasping that they are sounds of speech. Speech sounds interest us because they bear meaning and communicate linguistic information.

On a traditional account of perceiving spoken language, we hear certain sounds and then grasp their meanings. We auditorily perceive sounds, but we *understand* their meanings. On this account, hearing speech sounds is just like hearing non-speech sounds, except in its effects upon the understanding. Speech sounds cause us to grasp meanings with which they are contingently associated.

Nevertheless, hearing speech in a language you understand differs from hearing speech you do not understand. The difference is not just that in one case, but not in the other, you associate meanings with the sounds you hear. The difference is unlike that between seeing written words you understand and seeing those you do not. The visual experience of the shapes and spacing of letters and words does not change dramatically when you understand them. However, the sounds themselves differ in auditory appearance once you learn a spoken language. You hear pauses, word boundaries, and subtle distinctions in vowel and consonant sounds that you previously did not hear. Understanding a spoken language makes a distinctive difference to the phenomenology of hearing speech sounds.

John McDowell has claimed that understanding a language makes possible the experience of sounds as publicly *meaningful* (1998a, 1998b; see discussion in Smith, Chapter 9). Hearing meaningfulness implies a difference in auditory experience between listening to speech in a language you know and listening to speech in one you do not know. While it offers a richer account of the content of auditory experience in the case of speech perception, hearing meanings does not explain why we experience sounds to have different acoustic qualities once we hear them as meaningful (since meanings lack acoustic characteristics). It also invites us to ask the challenging question: What is the *auditory* experience of meaning like?

Barry Smith (Chapter 9) advocates a more conservative response than McDowell to the traditional account of the roles of audition and the understanding in hearing spoken language. Smith suggests that, while we understand but do not hear meanings, we do hear more than just the sounds of speech. Smith argues that we are auditorily aware of the *voices* of individual speakers, in addition to the apparent sounds of speech. Awareness of voices, rather than hearing meanings, accounts for our sense of communicative contact with verbal language users. Two features of Smith's account are noteworthy. First, voices play a role similar to sound sources, considered above, in the content of audition. Both are among the things we auditorily experience thanks to hearing the sounds they produce. Smith's account of speech perception thus involves something like hearing sound sources. Second, though it avoids any mystery about the auditory experience of meanings, hearing as of a voice does not by itself account for changes in the experience of sounds and their attributes in a language we understand. Hearing voices can explain the difference that accrues thanks to hearing sounds as speech, but it cannot explain the further difference due to understanding that speech.

So, we might claim that the *contents* of speech perception experiences differ from those of hearing sounds in ordinary non-linguistic audition. Auditory contents when hearing speech might include, as we have seen, meanings or voices. On the other hand, we could attribute the phenomenological difference after learning a language simply to ascribing different audible or acoustic features to sounds themselves. Perhaps we acquire the capacity to hear subtle contrasts, pauses, and rhythms that make a phenomenological difference.

Each of the options considered above assumes that hearing speech involves hearing sounds. Speech sounds are meaningful; they are produced by voices; they have noteworthy audible qualities; but they are a type of *sound*.

Is hearing speech hearing sounds? Consider the phenomenology of speech perception. Many researchers have noted that audible speech seems, phenomenologically, like a neatly ordered, regimented sequence of distinguishable sound types known as *phonemes*, which make up distinct words organized into structured sentences. Phonemes are important in understanding the auditory perception of speech because they are the distinguishable, language-specific equivalence classes of sounds that make up the spoken words of a language. English contains approximately 40–44 phonemes, including /d/, /z/, /ʃ/ (<sh>), and /ə/. Languages differ in what sounds they distinguish as different phonemes and in what sounds they count as allophones (variants) of a single phoneme. Spanish, for instance, does not distinguish /s/ from /z/. Perhaps, then, learning a spoken language requires the capacity to hear and distinguish the phonemes that make up its words, along with their specific audible characteristics.

A great source of dispute in this area stems from a vast body of empirical research that suggests a substantial divergence between the experienced features of speech sounds and the actual features of acoustic signals. No consistent cues recognizable in an acoustic signal, such as frequency or amplitude patterns, straightforwardly determine what one hears as a given phoneme or word (see Mole, Chapter 10, and Remez and Trout, Chapter 11, for further discussion). In particular, the acoustic features that correspond to a given phoneme depend upon the phonemic context. Both prior and subsequent phonemes impact the acoustic signature for a specific phoneme occurrence. Furthermore, acoustically, different speakers differ dramatically. The main philosophical lesson is that the manifest image of speech and the scientific image of sounds appear sharply disconnected.

One response is anti-realism about speech sounds. Georges Rey (2007, 2008), for instance, argues that phonemes and other linguistic entities are mere intentional objects that commonly lack physical instances. Smith draws a similar lesson from the divergence between phenomenology and acoustics. Smith contends that while acoustic signals do not contain the linguistic sounds or structures we seem to hear, we do manage to communicate by speaking. Communication, he claims, does not require the existence of speech sounds in the world, but only requires that the world seems to contain linguistic entities.

Notice that, unless eliminativism or anti-realism is true of sounds in general, anti-realism about the objects of speech perception implies that the objects of speech perception differ from those of ordinary audition.

Some have argued explicitly that the objects of speech perception differ from those of ordinary non-linguistic audition, since, given the empirical evidence, hearing speech is not hearing sounds. For example, Liberman (see 1996) famously and influentially argues that the objects of speech perception are *intended motor commands*, since aspects of the *production* of speech, such as the articulatory gestures used to generate it, do have affinities (if not complete correspondence) with and perhaps do predict experienced phonemes. Liberman even argues that since it targets intended motor commands rather than sounds, perceiving speech invokes a dedicated perceptual module distinct from audition. Perceiving speech and perceiving sounds on this view require different perceptual modalities.

Mole (Chapter 10) is critical of Liberman's motor theory. Mole argues that it is unclear whether the motor theory is supposed to provide an account of what we experience in perceiving speech or of what is represented by subpersonal structures implicated in perceiving speech. In the former case, Mole argues, it is phenomenologically implausible. In the latter case, it is unwarranted.

Worse, it is excessively demanding and thus untenable as a claim about what is represented by the mechanisms of speech perception.

We would like to note that the empirically grounded argument that hearing speech is not hearing sounds is unsound unless sounds straightforwardly can be identified with or are determined by underlying acoustic features. There are good reasons to doubt this. Neither the sound of a car driving on a gravel road, nor the sound of wood striking wood, for example, corresponds to a simple or straightforward feature recognizable on the surface of the acoustic signal. Each is highly complex and probably requires mentioning features of its source to make its individuation intelligible. Good reasons suggest that even the qualities of pitch, timbre, and loudness lack straightforward physical correlates. Hearing speech might not be distinctive, after all.

Fowler's (1986) *direct realist* account of speech perception attempts to capture the importance of articulatory movements of the mouth and vocal tract to speech perception while arguing that speech perception is a form of ordinary, environmentally situated audition. All audition, Fowler claims, is a matter of using acoustic information to find out about things and happenings in one's environment. If ordinary audition involves awareness of sound sources, and if, counter to a very naïve physicalism, we should not expect to match features of heard sounds with straightforward acoustic features, then despite the empirical results about speech, the objects of speech perception and the objects of non-linguistic audition might belong to a common kind.

Remez and Trout (Chapter 11) draw a stronger lesson from the discoveries of speech perception science during the past century. Remez and Trout argue that no reductive account of the objects of speech perception is compatible with the empirical evidence. Thus, the objects of speech perception are reducible neither to sounds nor to intended gestures nor to articulatory movements. Instead, according to their *homeostatic properties* account, speech perception depends upon properties that are diagnostic of, but not identifiable with, particular speech sounds. The diagnostic features for speech perception might be highly theoretical and closed to introspection. To discover what such features are requires examining the processes that underlie the perception and recognition of speech.

Remez and Trout argue that the case of speech illustrates in a particularly poignant way a more general lesson. The use of introspection and phenomenological considerations in theorizing about perception and its objects requires independent justification that it has not received. They argue that without a justification, nothing of use to scientific psychology comes from examining phenomenology. For instance, they claim that considering what the experience of speech perception is like distracts from the scientific task of explaining speech perception and linguistic understanding. Remez and Trout thus take the lessons of speech perception to warrant a general warning against relying on any methodology that uses phenomenology to discern the structure either of perceptual content or of perceptual processes.

It remains to be settled whether speech perception has special content, or has special objects other than sounds, or invokes special perceptual systems. This is fertile territory not only for conceptually sophisticated empirical work but also for philosophical and theoretical contributions. It is, however, uncontroversial that speech sounds are particularly salient and significant for humans, and that we enjoy special sensitivity to speech sounds. Human infants at a very early age distinguish speech from non-speech and show greater interest in speech sounds than in similarly complex non-speech sounds (Vouloumanos and Werker 2007). The capacities that support this interest remain to be characterized and explained.

5. Concluding Remarks

The issues we have discussed form the heart of the philosophy of sounds and auditory perception, as we understand it. The main debates concern the ontological nature of sounds; the locations of sounds; the characterization of spatial audition; whether and how we hear sound sources in addition to sounds; the nature of musical listening; and the primary objects of speech perception. We have pointed out where each question, in addition to being interesting in itself, promises to impact theorizing about perception more broadly.

We would like to conclude with a remark about a point of concern. Remez and Trout's (Chapter 11) warning about phenomenology calls into question the phenomenological constraint upon theorizing about auditory perception from Section 2.2. We claimed that phenomenology is prima-facie relevant to theorizing about the content of auditory perceptual experience. We claimed that one way to capture this is in terms of the veridicality conditions for auditory experiences, the appropriateness of which we discern in part phenomenologically. But other worries recently have been expressed about the reliability of phenomenological reports (see, e.g., Schwitzgebel 2008; Jack and Roepstorff 2003; Roepstorff and Jack 2004), and it is now commonly accepted that introspecting phenomenology is an imperfect guide to understanding perception. Phenomenological reports are influenced by numerous factors beyond just what experience presents itself as being like for its subject. Perhaps, therefore, we should cast out introspection as a way to understand the contents and objects of audition.

We should, however, distinguish using introspection as a guide to the structure of the mind and mental processes from using introspection as a guide to how the objects of experience appear. We believe that, in the second sense, introspective phenomenology is relevant to theorizing about perception. Illusions and hallucinations are differences between how things appear and how things are. How things appear is a matter of phenomenology. How things appear impacts what we believe and what we do. What we believe and do arguably are matters impacted by perceptual content. This grounds a case for the prima-facie relevance of phenomenology to philosophical questions about the content of experience. Phenomenology in this sense also figures centrally in psychological research on perception. Theories of vision aim in part to explain why things look they way they do (see Pylyshyn 2003: ch. 1). Theories of audition similarly aim to explain why things sound the way they do. That includes explaining how audition presents things as being and why auditory experience is organized as it is (Bregman 2005). The data for these theories thus are partly introspective—they includes first-person descriptions of what can be seen or heard, and the way those things look or sound. Reports of phenomenology are data that must be explained by a psychological theory, even if only part of the explanation is that experiences have features accessible to and reportable by the subject.

What about the reliability of introspective or phenomenological methods? We should distinguish unreflective introspection from careful phenomenological inquiry. We also should not presuppose that phenomenological descriptions are obvious or self-evident. As with other data, such descriptions may be revised or rejected in the light of subsequent thinking. Done with care, however, introspection may lead to interesting insights about what perceptual experience is like and what perceptual theorizing must explain (see, e.g., the essays in Noë 2002, 2007).

Introspection, in conjunction with the kinds of philosophical methods used by the contributors to this volume, can help make clear just what any satisfactory account of auditory perception and experience must address. Many of the questions raised in the chapters that follow, such as those that concern the content of auditory experience, the experience of music, and the perception of speech, will not, however, be resolved by introspection alone. Confronting and solving these problems will require whatever insights we can glean from psychological theorizing and from philosophy. But until good reasons show that subjectively accessible features of experiences are irrelevant to psychological theorizing—about, for instance, concepts and action—we continue to maintain the minimal thesis that, all else equal, theorizing about perceptual content should respect phenomenology.

References

Batty, C. (2007). *Lessons in Smelling: Essays on Olfactory Perception*. Ph.D. thesis. Cambridge, Mass.: MIT.

Blauert, J. (1997). *Spatial Hearing: The Psychophysics of Human Sound Localization*. Cambridge, Mass.: MIT Press.

Bregman, A. S. (1990). *Auditory Scene Analysis: The Perceptual Organization of Sound*. Cambridge, Mass.: MIT Press.

—— (2005). 'Auditory Scene Analysis and the Role of Phenomenology in Experimental Psychology'. *Canadian Psychology*, 46: 32–40.

Casati, R. and Dokic, J. (1994). *La Philosopie du Son*. Nîmes: Chambon.

—— —— (2005). 'Sounds', in E. N. Zalta (ed.), *The Stanford Encyclopedia of Philosophy* (Fall 2005 edition), available at <http://plato.stanford.edu/archives/fall2005/entries/sounds/>.

Crane, T. (2009). 'Intentionalism', in B. McLaughlin, A. Beckermann, and S. Walter (eds.), *Oxford Handbook of Philosophy of Mind*. Oxford: Oxford University Press.

Evans, G. (1980). 'Things Without the Mind—A Commentary upon Chapter Two of Strawson's *Individuals*', in Z. van Straaten (ed.), *Philosophical Subjects: Essays Presented to P. F. Strawson*. Oxford: Clarendon Press. Reprinted in G. Evans, *Collected Papers*. Oxford: Clarendon Press, 1985.

Fowler, C. A. (1986). 'An Event Approach to the Study of Speech Perception from a Direct-Realist Perspective'. *Journal of Phonetics*, 14: 3–28.

Gallagher, S. (2005). *How the Body Shapes the Mind*. New York: Oxford University Press.

Hamilton, A. (2007). *Aesthetics and Music*. London: Continuum.

Handel, S. (1993). *Listening: An Introduction to the Perception of Auditory Events*. Cambridge, Mass.: MIT Press.

—— (1995). 'Timbre Perception and Auditory Object Identification', in B. C. Moore (ed.), *Hearing*. San Diego, Calif.: Academic Press, 425–61.

Jack, A. and Roepstorff, A. (eds.). (2003). *Trusting the Subject: The Use of Introspective Evidence in Cognitive Science*, volume 1. Charlottesville, VA: Imprint Academic.

Liberman, A. M. (1996). *Speech: A Special Code*. Cambridge, Mass.: MIT Press.

Lycan, W. (2000). 'The Slighting of Smell', in N. Bhushan and Rosenfeld (eds.), *Of Minds and Molecules: New Philosophical Perspectives on Chemistry*. Oxford: Oxford University Press, 273–89.

McAdams, S. and Bigand, E. (eds.) (1993). *Thinking in Sound: The Cognitive Psychology of Human Audition*. New York: Oxford University Press.
McDowell, J. (1998a). *Meaning, Knowledge, and Reality*. Cambridge, Mass.: Harvard University Press.
——(1998b). *Mind, Value, and Reality*. Cambridge, Mass.: Harvard University Press.
Maclachlan, D. L. C. (1989). *Philosophy of Perception*. Englewood Cliffs, NJ: Prentice Hall.
Malpas, R. M. P. (1965). 'The Location of Sound', in R. J. Butler (ed.), *Analytical Philosophy*, second series. Oxford: Basil Blackwell, 131–44.
Martin, M. G. F. (1992). 'Sight and Touch', in T. Crane (ed.), *The Contents of Experience*. Cambridge: Cambridge University Press.
——(2006). 'On Being Alienated', in T. S. Gendler and J. Hawthorne (eds.), *Perceptual Experience*. New York: Oxford University Press, 354–410.
Matthen, M. (forthcoming). 'Auditory Objects'. *Review of Philosophy and Psychology*, 7.
Neuhoff, J. (2004). *Ecological Psychoacoustics*. London: Academic Press.
Noë, A. (ed.). (2002). *Is the Visual World a Grand Illusion?* Charlottesville, VA: Imprint Academic.
——(2004). *Action in Perception*. Cambridge, Mass.: MIT Press.
——(ed.). (2007). *Phenomenology and the Cognitive Sciences* (special issue on Dennett's heterophenomenology), 6: 1–270.
Nudds, M. (2001). 'Experiencing the Production of Sounds'. *European Journal of Philosophy*, 9: 210–29.
O'Callaghan, C. (2007). *Sounds: A Philosophical Theory*. Oxford: Oxford University Press.
——(2008). 'Object Perception: Vision and Audition'. *Philosophy Compass*, 3: 803–29.
——(forthcoming). 'Perceiving the Locations of Sounds'. *Review of Philosophy and Psychology*, 1.
O'Shaughnessy, B. (1957). 'The Location of Sound'. *Mind*, 66: 471–90.
——(1989). 'The Sense of Touch'. *Australasian Journal of Philosophy*, 69: 37–58.
Pasnau, R. (1999). 'What is Sound?' *Philosophical Quarterly*, 49: 309–24.
Pylyshyn, Z. (2003). *Seeing and Visualizing*. Cambridge, Mass.: MIT Press.
Rey, G. (2007). 'Externalism and Inexistence in Early Content', in R. Schantz (ed.), *Prospects for Meaning*, volume 3 of *Current Issues in Theoretical Philosophy*. New York: de Gruyter.
——(2008). 'In Defense of Folieism: Replies to Critics'. *Croatian Journal of Philosophy*, 8: 177–202.
Roepstorff, A. and Jack, A. (eds.) (2004). *Trusting the Subject: The Use of Introspective Evidence in Cognitive Science*, volume 2. Charlottesville, VA: Imprint Academic.
Scholl, B. J. (2001). 'Objects and Attention: The State of the Art'. *Cognition*, 80: 1–46.
Schwitzgebel, E. (2008). 'The Unreliability of Naive Introspection'. *Philosophical Review*, 117: 245–73.
Scott, M. (2001). 'Tactual Perception'. *Australasian Journal of Philosophy*, 79: 149–60.
Scruton, R. (1997). *The Aesthetics of Music*. Oxford: Oxford University Press.

Smith, A. D. (2002). *The Problem of Perception*. Cambridge, Mass.: Harvard University Press.
Smith, B. C. (ed.) (2007). *Questions of Taste: The Philosophy of Wine*. Oxford: Oxford University Press.
Strawson, P. F. (1959). *Individuals*. New York: Routledge.
Tye, M. (2000). *Consciousness, Color, and Content*. Cambridge, Mass.: MIT Press.
Vouloumanos, A. and Werker, J. F. (2007). 'Listening to Language at Birth: Evidence for a Bias for Speech in Neonates'. *Developmental Science*, 10: 159–64.

2

Sounds and Events[1]

CASEY O'CALLAGHAN

I argue that sounds are best conceived not as pressure waves that travel through a medium, nor as physical properties of the objects ordinarily thought to be the sources of sounds, but rather as events of a certain kind. Sounds are particular events in which a surrounding medium is disturbed or set into wavelike motion by the activities of a body or interacting bodies. This Event View of sounds provides a unified perceptual account of several pervasive sound phenomena, including transmission through barriers, constructive and destructive interference, and echoes.

1. What is a Sound?

Sounds are public objects of auditory perception. When a car starts it makes a sound; when hands clap the result is a sound. Sounds are what we hear during episodes of genuine hearing. Sounds have properties such as pitch, timbre, and loudness. But this tells us little about what sort of thing a sound is—which metaphysical category it belongs to. This is the question I wish to answer.

---

[1] This chapter was presented as a paper at the University of London conference on Sounds in 2004, at which this collection was conceived. It states concisely some central components of my account of sounds, which I went on to develop in greater detail and to expand upon in O'Callaghan (2007). I received helpful feedback on this version of the chapter from a number of individuals and audiences. In particular, I thank Paul Benacerraf, John Burgess, Scott Jenkins, Mark Johnston, Simon Keller, Sean Kelly, the late David Lewis, Matt Nudds, Robert Pasnau, Gideon Rosen, Roger Scruton, and Jeff Speaks for very helpful discussion and comments. I also thank audiences at Princeton University, University of California Santa Cruz, Auburn University, University of St Andrews, and Bates College. Finally, I thank Roberto Casati for his valuable commentary at the London conference.

2. Three Theories of Sound

Locke held that sounds are properties of bodies. More specifically, he held that sounds are secondary qualities: sensible qualities possessed by bodies in virtue of the 'size, figure, number, and motion' of their parts, but nonetheless distinct from these primary attributes (*Essay*, II.8). Robert Pasnau (1999, 2000) has recently proposed an account according to which sounds are physical properties of ordinary external objects. On what I will call the *Property View*, an object 'has' or 'possesses' a sound when it vibrates at a particular frequency and amplitude. Pasnau claims that sounds are properties of objects, though he reduces sound to the primary quality that is the categorical base of Locke's power, i.e., that of vibration or motion of a particular sort.

The received view of auditory scientists and physicists is quite different. It holds that a sound is a disturbance that moves through a medium such as air or water as a longitudinal compression wave. Vibrating objects produce sounds, but sounds themselves are waves. When we hear sounds, we do not immediately hear bodies or properties of bodies; we hear the pattern of pressure differences that constitutes a wave disturbance in the surrounding medium.

The common interpretation of Aristotle is that he held a very similar view. *De Anima* (II.8, 420b10) says that 'sound is a particular movement of air', which seems to indicate that Aristotle held a version of the received view, or as I will call it, the *Wave View*. We can, however, take our interpretative cues from other passages in the same chapter and arrive at a view that has certain advantages over the other two theories and will be at the core of the alternative I will develop. At 420b13, Aristotle says that 'everything which makes a sound does so because something strikes something else in something else again, and this last is air'. So, a striking causes or makes a sound when it happens in air. The sound itself is a movement. But the sound need not be the motion *of the air itself*. Instead, it may be the event of that medium's being disturbed or moved. The idea is to treat 'movement' as the nominalization of a transitive verb and focus on constructions like '*x* moves *y*' instead of '*y* is moving'. '*For sound is the movement of that which can be moved* in the way in which things rebound from smooth surfaces when someone strikes them' (420a20) means that sound is the air's being disturbed by the motion of an object. A sound is not motion, but the act of one thing moving another. This is not the Wave

View that most attribute to Aristotle, but the beginnings of an *Event View* of sound.

According to the Event View I propose, sounds are particular events of a certain kind. They are events in which a moving object disturbs a surrounding medium and sets it moving. The strikings and crashings are not the sounds, but are the causes of sounds. The waves in the medium are not the sounds themselves, but are the effects of sounds. Sounds so conceived possess the properties we hear sounds as possessing: pitch, timbre, loudness, duration, and as we shall see, spatial location. When all goes well in ordinary auditory perception, we hear sounds much as they are.

3. Locatedness and the Wave View

According to the Wave View, sounds are waves. A particular sound is a train of waves that is generated by a disturbance and that moves through the surrounding medium. But this is not how things seem. When we hear a sound, we hear it to be located at some distance in a particular direction. In ordinary cases, sounds themselves, not merely their sources, seem to be located distally. Auditory scientists call this phenomenon '*externalization*'.[2] Sounds are not perceived, however, to travel through the air as waves do. They are heard to be roughly where the events that cause them take place. A police tip sheet entitled, 'How to Be a Good Witness' instructs individuals to 'Look in the direction of the sound—make a mental note of persons or vehicles in that area' (Kershaw 2002). If auditory experience is not systematically illusory with respect to the perceived locations of sounds, then sounds are not waves, since they are not perceived to be where the waves are.[3]

The argument depends on a phenomenological claim. Sounds are perceived to have more or less determinate locations. When we hear a clock ticking, the sound seems to be 'over there' by the clock; voices are heard to be in the neighborhood of speakers' heads and torsos; when a door slams in another part of the house, we know at least roughly where the accompanying racket takes place. I mean that we experience sounds, in a wide range of cases, to be located at a distance from us in a particular direction. When we do not, as when a sound seems to fill a room or engulf us, the sound is perceived to be

---

[2] Gelfand (1998: 374) refers to this phenomenon as 'extracranial localization': 'Sounds heard in a sound field seem to be localized in the environment'. See also Blauert (1997).

[3] Pasnau (1999) argues that spatial auditory experience conflicts with the Wave View of sound unless hearing is illusory.

all around, or at least in a larger portion of the surrounding space. Hearing a sound located in the head when listening to earphones is another sort of sound location perception, albeit a touch odd.[4]

Often, however, it is natural to describe sounds as coming from their sources. We ask where the buzzing sound is coming from and wonder whether the sound of the cougar came from ahead or behind. If sounds seem to *come from* particular places, in a spatial sense of 'coming from', then locatedness as I have characterized it does not accurately capture the phenomenology of auditory spatial perception.

How are we to take talk of sounds' being heard to come from a location? Do sounds seem to come from locations outside the head, or do they seem to have relatively stable locations outside the head? It might be that sounds are heard to come from a particular place by being heard first at that place, and then at successively closer intermediate locations. This is not the case with ordinary hearing. Sounds are not heard to travel through the air as scientists have taught us that waves do. Imagine a scenario in which engineers have rigged a surround-sound speaker system to produce a sound that seems to be generated by a bell across the room. This sound subsequently seems to speed through the air toward you and to enter your head like an auditory missile. This would indeed be a strange experience, one unlike our ordinary experiences of sounds, which present them as stationary relative to the objects and events that are their sources.

Perhaps sounds are heard to *be* nearby, but to have *come from* a particular place, much as a breeze seems to have come from a certain direction. But feeling a breeze is like listening with earphones: direction without distance. Earphone listening differs from ordinary hearing not just in where sounds seem to come from, but also in where sounds are heard to be. Imagine feeling *where the fan is* by feeling its breeze. Since sounds seem to come from sources in a sense that includes distance as well as direction, and not travel, the best sense to make of sounds' seeming to come from particular locations is that they have *causal sources* in those locations.

Given the phenomenological facts, the degree to which auditory location perception is illusory or misleading should follow from a theory of sound. No theory should make the fact of location perception a wholesale illusion, though individual instances of location perception might mislead about the actual locations of sounds. Thus, I might correctly hear a stereo speaker's

---

[4] Gelfand (1998: 374) refers to this phenomenon as 'intracranial lateralization': 'Sounds presented through a pair of earphones are perceived to come from within the head, and their source appears to be lateralized along a plane between the two ears'.

sound as located at the speaker itself; but I might undergo an illusion of hearing the sound to be located five feet to the right of that speaker. In both cases, I correctly perceive the sound to have a location, but the experience is inaccurate in the second instance. Occasionally, sound location perception is to some degree anomalous, as when sound seems to be all around in a reverberant room, when it seems to be in the head during headphone listening, or when the sound seems to be behind a jet plane overhead. Whether and when a sound can literally *fill* a reverberant room, be *inside* the head of a subject, or be *behind* an airplane will depend upon one's theory.

The phenomenon of locatedness spells prima-facie trouble for the Wave View. Sound waves pervade a medium and move through it at speeds determined by the density and elasticity of the medium. Yet we neither hear sounds as air sloshing around the room nor as moving roughly 340 meters through the air each second. Sounds are perceived to be relatively *stationary* with respect to their sources. The sound of a moving train seems to move only insofar as the train itself moves. When the train stops moving, so does its sound.

The trouble for the Wave View is serious. Since sounds are heard as having stable distal locations, either the sound is not identical with the sound waves, or we misperceive sounds in one important respect. If the sound is identical with the sound waves, the situation is not that we sometimes misperceive sounds, as when a sound ahead is heard to be behind; rather, we *systematically* misperceive the locations of sounds. That is, we hear the locations of all sounds incorrectly since we never hear a sound to move just as wavefronts do. Since sounds are among the things we hear, we should take the phenomenology of auditory experience seriously when theorizing about what sounds are. If the phenomenon of locatedness is not systematic misperception, then sounds are not sound waves.

The Wave theorist might reply,

The immediate objects of auditory perception—what we hear—are waves. Sounds just are waves. Waves and their properties are the causes of perceptions of pitch, loudness, and duration; however, we hear these qualities to be located at the place where the waves originate, i.e., at their source. Sounds seem to be where their sources are, and to this extent, auditory perception is illusory. But this illusion is a beneficial one, given our interest in sound sources as constituents of the environment. It is no surprise that we hear sounds to be located where distal objects and events are.

The Wave theorist's response avoids the conclusion that sounds are not identical with waves by accepting that we are subject to wholesale illusion in one salient aspect of auditory experience. The strategy is to assuage concern about the location illusion by providing another candidate for bearer of the spatial properties and by highlighting the illusion's potential benefits. Notice the tactic. By invoking the location of the *source*, the Wave theorist avoids assigning potentially problematic locations to *sounds*.

But an account that locates perceived instances of pitch, timbre, and loudness with their sounds is preferable, all else equal, to one that convicts auditory perception of systematic illusion about the locations of its objects. In part, the case against the Wave View depends on whether there exists an alternative view that captures the locatedness of sounds while matching or surpassing the Wave View's success at providing a unified explanation of other sound-related phenomena. Part of the task of this chapter is to develop such an alternative.

Before going forward, we must first consider: Can we eliminate the location illusion from the Wave theorist's account entirely? A final promising approach again rejects the phenomenological claim as it stands. Instead, it says that we hear sounds to have pitch, timbre, loudness, and duration, though not as having location. Rather, we hear ordinary events and objects as located and as the generators or sources of audible qualities that lack spatial properties entirely. We do not mistakenly perceive the locations of sounds, we simply fail to perceive their locations.

The Wave theorist avoids the dilemma by saying that sounds are not heard to have locations, they are heard to have located sources. The picture is this: Sounds are waves; waves have sources; sounds are heard to be generated by their sources, but not themselves to have locations; only sources are perceived to have locations. This description provides an account of the phenomenology that is consistent with the Wave View. Unfortunately for the Wave theorist, it fails. To see why it fails we need to consider just how audition furnishes perceptual information about the locations where sounds are generated.

Hearing provides information about ordinary objects and events around us—notably, information about where those things are and occur. (Try *not* to turn your head toward a book dropped behind you.) The response we are considering is that we hear objects and events as located by means of the sounds they generate. For the Wave theorist, the basic audible qualities are qualities of sounds, and sounds are waves. Thus, waves have the audible qualities. But we cannot *hear* just non-located audible qualities and located objects, *full stop*. This would amount to a precarious perceptual situation. How could hearing non-located qualities provide perceptual information about sound source locations?

One way is for locational information to be encoded temporally, for example, by time delays between waves reaching the ears. However, since we are auditorily *aware* of the locations of things and happenings—hearing is spatial—this information must be conveyed somehow in conscious perception. At the basic level of awareness, audition presents just complexes of pitch and timbre with loudness and duration, so an auditory experience that conveys information about the locations of material objects and events must do so by means of one's awareness of these basic attributes. Temporally encoded location information is manifested through one's experience of pitch, timbre, and loudness.

For an experience of the audible qualities to be an auditory experience of location, the audible qualities must themselves bear spatial information. Given that, as I have argued, sounds and their audible qualities do not auditorily seem to come from particular locations in a sense that involves travel or arrival, auditory awareness of location must occur thanks to an awareness of located audible qualities. Sounds, the bearers of audible qualities, must appear to occupy stable distal locations if we are to learn of those locations through auditory experience.

A distinction can thus be drawn between hearing sounds themselves as located and perceiving information about the locations of material objects, stuffs, and events in the environment by means of audition. Given that we learn the locations of ordinary objects and events in audition, the question is whether the latter would be possible without the former. Since sounds seem to come from their sources only in a causal sense, and since auditory awareness of location must occur by means of awareness of audible qualities, hearing sounds and their qualities as located is required in order to perceive or form judgments about the locations of material objects and events through audition. Sounds are heard to have locations, by means of which they provide perceptual information about the locations of their sources. If the Wave View is correct, the location illusion remains.

If the phenomenological claim is an accurate description of the experience of sounds, and if it is true that in order to perceive the locations of sound sources, audible qualities and sounds themselves must be perceived as located, then either the Wave View attributes widespread illusion to auditory perception or the Wave View is false and sounds are not simply waves in a medium. Short of accepting and explaining the illusion, the Wave theorist's best strategy is to impugn my description of the phenomenology. She should say that sources, not sounds themselves, are heard as located. This requires rejecting the argument that in order to hear sound sources as located, sounds must be heard as located. I know no simple route around the dilemma for the Wave theorist shy about claiming that we fail to perceive, or that we systematically misperceive, the locations of sounds.

I am convinced that the phenomenological claim is correct as it stands and that sources are heard as located only if sounds are. So I need to avoid this difficulty. My theory must not imply that sounds move through the air. The Property View is tailor-made to capture the phenomenology of locatedness. It, however, falls to a separate objection.

4. The Argument from Vacuums

The Property View says that sounds are properties of things like bells, tuning forks, and whistles; more specifically, sounds are the vibrations of material objects. The view entails that sounds are roughly where we perceive them to be. Unfortunately, the Property View also entails that sounds can exist in the absence of a transmitting medium. That is, sounds can exist in a vacuum (just as things can have colors in the dark), since all that is required for an object to have a sound is that it vibrate in the right way.[5] Nevertheless, we have good reasons to believe that the existence of a sound requires a medium. If there can be no sounds in vacuums, the Property View is false.

In Berkeley's first dialogue between Hylas and Philonous, Hylas argues against the Property View in favor of the Wave View by deploying the *Argument from Vacuums*. It begins with the premise that a bell struck in water or air makes a sound, but in a vacuum it does not. Hylas concludes that sound must be in the medium.

PHILONOUS. Then as to sounds, what must we think of them: are they accidents really inherent in external bodies, or not?

HYLAS. That they inhere not in the sonorous bodies, is plain from hence; because a bell struck in the exhausted receiver of an air-pump, sends forth no sound. The air therefore must be thought the subject of sound. [The sound which exists without us] is merely a vibrative or undulatory motion in the air. (Berkeley 1975: 171–2; quoted in Pasnau 1999: 321)

The argument is:

1. A bell struck in a vacuum makes no sound.
2. So, sound does not exist in the absence of air.
3. So, air is the subject of sound (i.e., the Wave View is true).
4. The Property View is false.

Notice a few things. If a bell struck in a vacuum makes no sound, then sound does not exist in the absence of air (or some other medium) and the Property

---

[5] It also entails that there does not exist a *causal* relation between a source and its sound.

View is false. But it does not follow that the Wave View is true. Air might be required for the existence of sound without itself being the subject of sound. Even if its first premise is true, the Argument from Vacuums does not establish the truth of the Wave View. Room exists for an alternative theory of sounds according to which no sounds occur in vacuums.

Furthermore, the first premise must be established if it can be used against the Property View. Why say there are no sounds in vacuums? Hylas baldly assumes there are not. We would like to have some reason, preferably independent of an explicit theoretical commitment, for denying (or affirming) that sounds exist in vacuums.

A first pass: When the bell is struck in a vacuum we know there is no sound because none can be heard by any ordinary creatures; if a medium is added, we can hear it, so the sound must require the medium. Problem: The fact that no sounds are ever heard in the absence of a medium shows only that a medium is required for there to be veridical perception of sounds. It does not show that a medium is necessary for there to be a sound.[6] This is the Property theorist's wedge in the Argument from Vacuums. Suppose one strikes a bell in a vacuum chamber containing a (hypothetical) perceiver. The perceiver can hear nothing. Our problem is that without a theory of sounds we are unable to confirm whether or not there is a sound. Barring a declared theoretical commitment, how do we decide if the bell makes a sound?

Talk of vacuums and sounds might end here, until we have chosen among competing accounts of the metaphysics of sounds. We might, on the other hand, simply bar any view that permits sounds in vacuums, on the grounds that it is too much at odds with common sense. Neither is required; good reasons suggest that sounds cannot exist in vacuums, whether we can confirm it or not.

Perhaps because the bell struck in a vacuum is not a possible object of auditory experience, it does not make a sound? Though I see no good reason to deny that there are sounds beyond the ken of perception, this argument gets us closer to what we are looking for. A sound, if anything, is the bearer of the properties of pitch, timbre, and loudness. Suppose we could establish that there is neither pitch, nor timbre, nor loudness when the bell is struck in the vacuum. We could then reasonably conclude that there is no sound. The bell struck in a vacuum has no sound because it has none of the qualities necessary for the existence of a sound.

---

[6] Berkeley, of course, had reason enough to conclude that there are no sounds in vacuums, since he accepted that nothing exists unperceived.

The sound of a bell seems to have different qualities when the bell is struck in air and water, and different ones yet in helium and liquid mercury. When the very same striking event occurs in a vacuum, it is inaudible. If a sound exists in a vacuum, it must have some definite pitch, timbre, and loudness. What loudness, for example, does it have? Does it have the loudness it would have been heard to have if it were surrounded by water? Does it have the loudness it would have been heard to have if it were surrounded by air? A decision here will be to a significant extent arbitrary and will not reflect the relevant ways in which the loudness of a sound depends upon the medium in which it is generated.

The Property theorist might hope that ideal or standard conditions for perceiving the 'true sounds' of things can be formulated as they can for colors.[7] If so, the pitch, timbre, and loudness of a sound in a vacuum are just those it would appear to have in ideal or standard conditions. There is, however, an important disanalogy between colors and sounds. If colors depend on the reflective properties of a surface, then daylight or white light is *normatively significant* in a way that less-than-full-spectrum lighting is not. Once reflected, full-spectrum incident light carries information about how much light a surface reflects or emits at each wavelength across the entire visible spectrum. But the way in which daylight or white light might be counted as ideal for revealing the true colors of things finds no analog in sound. Neither air nor water nor helium does a substantially better job divulging the subtle vibrations of an object in the way that full-spectrum light reveals the reflective properties of a surface. If there is no ideal or normatively significant medium in which to hear the true sound of an object, and if the qualities of a sound depend upon the medium in which it is generated, then it is doubtful whether the object vibrating in a vacuum has a pitch, timbre, or loudness. It is therefore doubtful *whether there is any sound*.

Might we say that sounds are properties that objects have only when in the presence of a medium, and thereby save the Property View? The sound property assigned to the object must in this case depend upon the specific properties of the medium surrounding it in order to avoid the objection raised above. The sound differs when the medium differs. This is no longer the Property View. It is a relational view involving object and medium which is closer to the truth about sound, not the view that sound is an inherent property of objects.

The medium dependence of audible qualities shows that we are justified in drawing a stronger conclusion than that some necessary condition for sound

---

[7] Pasnau (1999: 322) appeals to just such a hope.

perception is missing in a vacuum. A necessary condition for there *to be a sound* is missing. If sounds do not occur in vacuums, the Property View is false.[8]

Moreover, we showed earlier that the Wave View entails systematic illusion about where sounds are. The Event View, however, is a natural alternative that attributes the right locations to sounds and does not entail that sounds exist in vacuums. The Event View is that particular sounds are events in which a medium is disturbed or set into wave-like motion by the movement of a body or interacting bodies. These *disturbance events* take place where we perceive sounds to be, and, because no medium is present to be affected, a vacuum contains no sounds.

5. The Event View

Particular sounds are events.[9] Sounds take time and involve change—at a minimum they begin, and usually they end. A number of qualitatively different stages or a single tone of uniform loudness may compose a sound. The sounds are the events in which a medium is disturbed or changed or set into motion in a wave-like way by the motions of bodies. Events such as collisions and vibrations of objects cause the sound events. Among the effects of sounds may be sound waves propagating through a medium and the auditory experiences of perceivers. Medium-disturbing events are what we hear to have particular pitch, timbre, loudness, and location. A body counts as in a state of sounding—making a noise—just in case it is in the midst of generating or causing a particular sound. Whenever there is a sound, there is a sounding.

The tuning fork struck in air is a simple case. The striking is an event that 'makes' a sound in virtue of the process by which the arms of the fork oscillate and create regular compressions and rarefactions in the surrounding air. Its creating the disturbance constitutes the tuning fork's sounding. The event of the tuning fork's disturbing the medium is the sound. We perceive

---

[8] In O'Callaghan (2007), I argue against property accounts, in general, on the grounds that they cannot deal adequately with the temporal characteristics of sounds. In particular, an account in which sounds are properties cannot easily capture the fact that sounds persist and survive changes to their audible attributes.

[9] My theory of sounds as events should be relatively insensitive to what particular theory of events is the correct one. Within reason, whatever events turn out to be, sounds should be events. Accordingly, I wish to work with the intuitive notion of events as particulars which take time and may or may not essentially involve change.

this sound event to have a constant pitch and timbre, a duration, a location, and diminishing loudness. In contrast, the sound of an owl's call is a more complex event characterized by a temporally extended pattern of changing pitch, timbre, and volume. Each call sounded is an event that consists in the disturbing by the owl's lungs and syrinx of the surrounding air in a given pattern. The tuning fork and the owl alike are recognizable by the sounds they create.

Auditory perception also makes us aware of events in our environment. We learn by audition how the furniture is arranged and when it is being moved. How is this possible if sounds themselves are the events that we hear? The Event View says that a sound is an event whose cause is the event heard to have or make the sound, and implies that the sound and its cause are in close spatio-temporal proximity, since we might treat the location of the disturbing event at a time as the surface of interaction between the object and the medium.[10] When we hear the sound of a glass breaking, that sound is an audible event constituted by the fracturing glass's affecting the air. The breaking of the glass causes the medium-affecting event that is the sound event. The medium-affecting event is near the breaking event, but the two do not occur in just the same space-time region. That the two sorts of events occur close to each other, however, does not sufficiently explain why we are aware of *sound-generating* events in auditory perception. A sound also carries qualitative information that can be used to identify its generating event after perceivers learn to associate the sound with the cause. The sound's pattern of pitch, timbre, loudness, and duration indicate that a glass has broken; the location of the sound points us in the direction of the mess.

So, there is the event of an object or substance setting a medium into periodic motion. This is a sound. The kind of motion depends on the form and makeup of the object or substance, what it does to disturb the medium, and the physical characteristics of the medium itself. The sound event has a location and a pattern of pitch, timbre, loudness, and duration. There are also the generating events that cause sounds and the objects that are said to make a sound in virtue of instances of their sounding.

Sounds are individuated along three primary dimensions: causal source, spatio-temporal continuity, and qualitative change. Intuition is sometimes silent, but we do have implicit in our practices principles for saying when sounds are the same or different. The Event View captures these principles.

---

[10] Cf. Bennett (1988: 12): 'The "location" of an event is its spatiotemporal location, i.e. where and when it occurs...A zone may be sizeless along one or more of its dimensions: ...some fill spatial volumes and presumably others occupy only planes, lines, even points'.

To count as the numerically same sound particular, a candidate must have the very same token causal source and be spatially and temporally continuous throughout its entire history. If either the causal source changes or there is a spatial or temporal discontinuity, we say that there are different relevant sound particulars—a temporally seamless transition from a trumpet playing B-flat to another trumpet playing the same note counts as involving two different sound tokens. The sense in which the sound from a single trumpet is different when it seamlessly goes from playing a B-flat to an A is that the trumpet's state of sounding is different. Perhaps different sound events correspond to these different states of sounding at the two times. Still, there is one sound event of which each note instance is a part, and in this sense both are parts of a single continuous sound. Such a sound might extend over considerable time and space and change greatly in its qualitative characteristics. At times it may be loud and high-pitched and at others it may be faint and low, but as long as it has the same causal source in terms of its generating event or object, and is spatio-temporally continuous, it may count as the very same sound particular. Often, however, an abrupt qualitative change signals distinct sounds.

Numerically distinct instances of sounds that fall under the same qualitative characterization are not *the very same sound* in any sense stronger than qualitative identity. Temporally discrete sounds from the same causal source and spatially separated sounds generated by different sources can at best be different instances of the same qualitative sound type. Philosophers sometimes do, however, speak of performances of songs and symphonies as events that are tokens of sound types, despite the fact that they need be neither temporally nor spatially continuous—they may incorporate periods of silence and multiple sources. We can say the same of bird calls. But these are complex events that involve patterns of individual sound events when they occur. Distinct sound particulars are arranged to comprise a whole that may require or allow for discontinuities of various kinds. The ontology of music and complex sound universals enjoys its own vast literature. What I want to point out is that the Event View is capable of capturing the ways in which we take sounds to be individuated. The principles I have mentioned may be disputed; but there is often obscurity about how events are to be individuated. The Event View, in that case, predicts—correctly—that there is a certain amount of obscurity and arbitrariness in our verdicts concerning how many sounds we have heard.

The Event View is a natural way to avoid the objections posed to the Wave View and Property View. Particular soundings have audible locations determined by where the medium-disturbing process occurs. Sounds, then, move through space in just those ways we expect them to, for example, when a train passes in the distance. The subject-directed missile-like sound does not ordinarily arise. The Event View also accounts for what we learn about sound from the Argument from Vacuums: we are justified in claiming that a medium is necessary for there to be a sound. Since a medium is required for there to be a medium disturbance, there is no sound in a vacuum. The Event theorist maintains that sounds are neither entirely in the surrounding medium nor simply properties of objects. If the arguments against the prevailing views are compelling, the Event View is a theoretically cogent solution.

Several lines of objection force elaboration of the Event View. The Event View provides for natural accounts of several phenomena that pose difficulties for any theory of sound.

6. Transmission

So far, I have said that a sound is an event of a medium's being disturbed or set into motion in a particular way by the activities of an object, body, or mass. But this seems too lenient, to allow too many sounds. Consider the following two forms of objection.

First form Suppose you are underwater and hear the sound of something that happens in the air above, say, the striking of a bell. The Event View seems to imply that there is a sound at the interface of the air and water since indeed there is a medium-affecting event there. The air, a mass or body, sets the water, a medium, into motion. This is phenomenologically inaccurate. We do not hear the sound to be at the surface of the water; we hear it to be above in the air.

Second form The preacher outside is loud. When I shut the window I do not hear him as well. The window muffles the sound. Nevertheless, the window also sets the medium inside the room into motion. According to the Event View, is the sound located at the windowpane? We do not hear it as being there—the sound still seems outside.

In both forms, sound waves generated in one medium pass into another kind of medium. The first describes travel across a single interface; the second involves travel through a solid barrier. In each, at the relevant interface—the air–water interface in the first case and the window–room interface in the second—the motion of a body disturbs the medium it adjoins. Yet since we do not ordinarily take ourselves to hear sounds at such places, intuition has it that no sound occurs at either the interface or the barrier. Must the Event theorist count these events as sounds?

The problem of transmission is not unique to the Event View. Each of the views canvassed faces a version of the objection. The Property View is in roughly the same straits as the Event View. The Property View implies that the sound is a property of the air mass in the first case and the windowpane in the second, since each vibrates at a particular frequency and amplitude. Even the Wave View, on which sounds are waves, faces a dilemma. What is the *source* of the sound? Is it the bell or the air–water interface? The preacher or the window? Each is in a sense the cause of the waves 'in the medium' in which the sound is heard. An acceptable version of the Wave View must acknowledge that we perceive locations in auditory perception, even if these are the locations of sound *sources*. Just as the Event theorist needs to say which events are the sounds, the Wave theorist must say which things count as sources of sounds. Though the problem is not unique to the Event View, the Event theorist owes an account of sounds and transmission.

The Event theorist's options are: (a) deny there is a sound where transmission occurs and explain why the Event View does not entail that there is; (b) accept that sounds accompany transmission events and reconcile this with the intuitive description of the experience. Contrary to first appearances, option (b) is somewhat attractive. Ultimately, however, this response with its burgeoning world of sounds is unsatisfactory. It strains the imagination to suppose that a multiplying of sounds occurs each time sound waves travel across an interface or through a barrier. Our accounting should be more sober.

Suppose we deny that a sound occurs when a 'new' medium is disturbed by a pre-existing sound wave. Option (a) suggests an attractive way to conceive of the perceptual situation. We say that the interface or barrier distorts our perception of the primary sound's location and qualities, not that we perceive a secondary sound with its own location and set of qualities that is caused by the primary sound. A single sound exists above the water or outside the window, but one may not have an ideal experience of that sound if impediments to perception intervene.

This picture is more accurate from a phenomenological standpoint. We have a perceptual bias toward the locations of sound-generating events of the everyday sort such as doors shutting and ocean waves breaking. We hear the sound created by the striking of the bell above water and the sound of the preacher proselytizing outside the window. Events of transmission occur when the waves from one sound event cause motion in an object or body that is passed on to another medium. We do not hear events of transmission or indeed anything at their locations when we hear a sound beyond an interface or through a barrier.

The language of this distinction points to a theoretical solution compatible with the Event View. To speak of a sound or of a sound wave as *generated* by a source implies that the sound or the wave is caused by and distinct from the event that brought it about. The idiom suggests that neither the sound nor the wave exists prior to an event of generation. In contrast, the idiom of *transmission* suggests the passing along of a wave disturbance that already exists. Indeed, the physics of sound wave generation differs from that of sound wave transmission. During generation, something which is not itself a sound wave produces a sound wave; during transmission, sound waves *travel through* an interface or barrier as sound waves. Sound events involve the active production of pressure waves, transmission events do not.

When a transmission event causes a medium disturbance of the sort that seemed to pose trouble for the Event theorist, that event depends for its existence upon a prior sound event. The distinction between events in which sound waves are introduced into an environment and those in which sound waves are transmitted is natural and based on the events' roles in a regular causal network. Being a sound is a matter partly of occupying a particular causal role. A central feature of the causal role distinguished by how we speak, one supported by the physics, is that sounds are events caused by generating events such as collisions, but are not caused simply by waves passing through boundaries and barriers. The medium-disturbing events that are the sounds are the events in which a wave disturbance is *introduced* into the environment by the activities of some material object, body, or mass. Medium-disturbing events in which a prior sound's waves are passed on or transmitted into a different medium are not in any ordinary sense sounds.

Suppose sound waves reach a barrier and induce vibrations in that object. The barrier might then itself generate a sound in addition to the sound whose waves induced the barrier's vibrations. This is not an ordinary case of sound wave transmission, however, and should be subsumed instead under *resonance*. Resonating is sounding since the resonating object actively disturbs the medium, and does not merely passively transmit existing sound waves.

This account appeases intuition. The problem of deciding which of multiple sounds we listen to when sound waves pass through an interface or barrier does not get off the ground. But the innocent picture according to which being a sound is entirely a matter of what happens near the surfaces of objects whose activities affect a medium is threatened. We must adopt a broader perspective that acknowledges the causal relations of several distinct kinds of events. This is not cause for alarm; nor is it a surprise, given the organization of sound-related experience. Sounds furnish us with awareness of sound-generating events, which are of paramount interest for what they tell us about the world. They tell us such things as how the furniture is arranged and when it is being moved. Transmission events, however, enjoy little utility beyond what we learn through their effects on how we perceive the primary sounds they occlude: when we perceive a sound as muffled, we learn that a barrier may intervene. Given our interest in ordinary events that take place among material bodies, along with how these events are related to sounds, it is no wonder that the primary disturbances should be distinguished by audible qualities.

7. Destructive and Constructive Interference

As commonly demonstrated in physics classrooms, sound waves *interfere* with each other. Suppose you are in an anechoic room in which two tuning forks tuned to E above middle C are simultaneously struck. As you move around the room, there are places from which you hear the sound to be soft and places from which you hear the sound to be loud; there are places from which you hear neither sound.

This phenomenon occurs because at any time the total pressure at a point in the room equals the algebraic sum of the pressures of all the sound waves at that point. It is therefore possible, when sound waves are out of phase with each other, for the total pressure at some point or in some area to remain constant while separate sound waves pass through that point or area simultaneously. A listener positioned at such a point hears nothing. When sound waves cancel, the interference is *destructive*. When the waves are completely in phase at a point, the total pressure varies with the sum of the components' amplitudes. The sound seems twice as loud as either tuning fork at these points thanks to *constructive* interference. Altering the phase or vibration characteristics of one of the tuning forks may result in *beating*, a periodic variation in perceived volume from a particular point.

Here is the problem. Take the example of complete destructive interference described above. The Wave theorist can explain that you hear no sound from where you stand because the pressure is constant at that point and hence there is no sound. Of course, there are still in a sense two sets of waves passing through that area, though their summed amplitude is zero. So, in a sense, there are two sounds at that point even though none is heard. The Wave theorist does not escape entirely. If, however, by 'the wave', we mean something that depends only on the total pressure at a point, there is no wave and no sound at the point in question. By contrast, the Event View implies that each tuning fork makes a sound even though you hear neither one from the point of interest. If sounds are not sound waves and the Event View is correct, then you hear no sound at all when there are two. Is the gap a fault line in the Event View?

Interference phenomena do not undermine the Event View. The interference arguments do show that waves carry information about sounds. The Event theorist should not deny this when he says that the sound is not identical with the waves. Waves can be involved in the process by means of which a sound is heard without the sound's just being the waves. The Event View provides an intuitive and compelling alternative to the standard account of destructive interference. The Event View says there are two sounds, two events of a disturbance being introduced into a medium. These disturbances travel as compression waves and may reach a perceiver, where they cause perceptions of the original sound event. Waves obey the principles of interference, and if no variations in pressure exist, no sounds are heard. Ordinarily, a lack of pressure variations indicates the absence of sounds and sound sources. Complete destructive interference resembles the absence of sounds because factors conspire to create nodes where the pressure does not vary. These factors include the spatial arrangement of the two sources, the frequency and amplitude at which the sources oscillate, and the temporal relations among the activities of the sources, i.e., the phase difference of the sources. A perceiver located at a node will hear neither sound, and may believe that no sounds occur. This does not entail that the room contains no sounds. The observer is simply unable to perceive the sounds because of her particular point of view.

That there are indeed two sounds can be confirmed in several ways. One can move to a point where one or the other sound is audible, move one or both of the sources so that the nodes are shifted, alter the phase difference in the vibrations of the two objects to remove nodes completely, or simply remove one of the sources to eliminate interference entirely. These exercises show that each tuning fork makes a sound that can be heard independently of the other in the right circumstances. Sometimes, however, another sound's presence can interfere with perceiving a given sound. Experience need not reveal from a particular vantage point all the surrounding environment's sounds. What we perceive from a very limited vantage point need not be the entire story about what sounds are around.

The case of constructive interference is very similar. Due to the spatial and temporal relations among events of sounding, a perceiver in the right location may experience multiple sources to have greater loudness than any single source present. This is again the result of the additive properties of sound waves. It is less surprising that the subject's loudness experience should increase in the presence of two sources than that it should decrease, as in destructive interference. Beating is perhaps less intuitively comprehensible, but is also an explicable result of how the source events are arranged in time and space, and of the subject's vantage point on these events.

8. Echoes

The phenomenon of an echo is familiar. You are at a fireworks display in an open field with a single brick building behind you. A colorful bomb's recognizable boom follows on the heels of its visual burst, but a moment later the boom's echo sounds at the brick wall behind the field. This phenomenon poses two potential problems for the Event View. First, is the echo a distinct sound event that occurs at the reflecting surface, or not? Though the echo seems distinct, the brick wall reflects sound waves and does not introduce a disturbance into the surrounding medium, so the Event View appears to have no sound to identify as the echo. Second, does the existence of echoes show that sounds themselves travel and can be re-encountered, and hence, that sounds are not the events I have suggested?

If echoes show that sounds are not events, then the Event View is false. So, the first question presupposes a negative answer to the second. I shall argue that echoes do not pose a problem for the Event View, and that once we have secured the correct conception of hearing an echo, the Event View has precisely the right kind of disturbance event on offer: the primary disturbance.

Matthew Nudds (2001: 221–2) has recently argued in the following way that sounds are not events.

1. Newton measured the speed of a sound by measuring the time it took for the sound to travel down a colonnade and back.
2. Hearing an echo is re-encountering a particular sound.
3. Events, unlike objects, cannot be re-encountered.
4. Therefore, sounds are object-like particulars and not events.

If this reasoning is cogent, an echo is a particular sound at a later stage of its continuous career, after it has been reflected.

Nudds also considers and rejects a two-part response: (a) Newton measured the speed of sound waves, but not of a sound; (b) an echo is a *distinct* sound whose qualities resemble the primary sound. Though I reject (b) for reasons I will soon discuss, I do accept (a). Claim (a) is a strong replacement for (1), from which (2) does not follow. But (2) does not follow even from the weaker (1*).

> 1*. Newton measured the speed of *sound waves* by measuring the time it took for him to hear an echo after hearing the primary sound.

Hearing an echo may not be re-encountering the same sound at a later stage of its career, even if we owe the episodes of hearing to the same sound waves. If (2) ought to be rejected on independent grounds, (1) is false. Since the conclusion that sounds are not events does not follow from the argument reconstructed with the uncontroversial (1*), the Event theorist can then provide an alternative account of echoes. What reason do we have to reject the claim that hearing an echo after its primary sound is re-encountering the same persisting sound particular?

Sounds are essentially extended in time—each sound has a beginning, a middle, and an end. Sounds are not wholly present at each moment at which they exist. Having a qualitative profile *over* time is central to the identity of a particular sound such as a spoken word or an owl's hoot. One kind of re-encounter we have with sound particulars includes those that occur at later stages 'during the completion' of the sound. This morning I heard the loud, high-pitched beginning of the local emergency siren's wail. I then descended into the silent basement for two minutes, after which I emerged to hear the nearly completed sound's fading low-pitched moan. This was a genuine re-encounter with the same particular sound at different times. I experienced a different part of the sound upon each hearing. Now, if Nudds is right, I can also hear a sound *in its entirety* during two (or more) distinct intervals during which it exists, and thereby re-encounter it. That would make sounds particulars that are *extended in time* and that *fail to be wholly present at any moment* at which they exist *and* that *can be experienced in their entirety at different stages of their continuous careers*. But a single particular cannot *continuously* exist throughout an interval of time during which it must begin and end—*entirely*—multiple times. Either sound duration perception—perceiving that sounds begin and end—is illusory, or the claim that sounds are persisting particulars that can be re-encountered in their entirety is false.

However, we perceive the durations of events that produce sounds, such as fingernails scraping across a blackboard, by perceiving the durations of sounds. We do not, in the first instance, perceive the durations of our experiences of sounds. But that is just what we must do if sound duration perception is an illusion that occurs in virtue of encounters with the spatial boundaries of passing sounds. How else could we perceive the durations of events such as blackboard scratchings if sounds did not have the durations we think they have? The trouble comes from thinking that a bout of echo perception is a re-encounter with the same sound particular later in its continuous career.

That claim finds little support in perceptual experience. Hearing an echo is unlike re-encountering a person you have met before; it is unlike glimpsing someone carrying home the vase you saw earlier in a store window, even though the echo has an equally rich qualitative signature. Echo perception does not bear the marks of object-recognition and identification that characterize the experience of material objects and continuants with relatively stable qualities. An echo seems to be distinct from its primary sound in a way that an object perceived at different times does not. Perhaps this is because particular sounds are often perceived to begin and end, or because one could not imagine continuously perceiving the entire sound as it traveled from source to wall and back. Whatever the explanation of this disanalogy, perceptual evidence fails to bolster the claim that sounds travel and can be heard again as echoes. In the face of the arguments, (2) should be rejected. If echo experiences are not re-encounters of the sort we have with objects, the conclusion that sounds are not events does not follow.

This brings us back to the first problem mentioned in this section. What event is the echo? Is an echo a distinct disturbance event that occurs at the reflecting surface? There are four reasons that together suggest it is not. First, awareness of an echo normally furnishes awareness of the event that made the sound. Hearing the echo of the firework's boom is a way of hearing the *explosion itself* again. A sound is always the sound of something happening. An echo experience, as well as that of a primary sound, can disclose those happenings. Second, we do not attribute dispositions to produce sounds with particular audible qualities to the reflecting surface. Third, and importantly, what occurs at the reflecting surface is not the introduction of a disturbance into the surrounding medium. An elastic collision between the surface and the medium occurs, causing the direction of wave propagation to change. Absent is the Event View's characteristic event: the original disturbance of a medium by the activity of a body. The reflecting body need not do anything but redirect pre-existing waves. Finally, an analogy with mirrors is compelling. Mirrors facilitate our seeing the very objects and events that occur in front of them, albeit with distortion of place. Likewise, reflecting surfaces allow us to hear the very sounds that occur in front of them, albeit with distortion of place *and time*, which results from the speed of sound waves. If the mirror analogy is correct, just as there are not distinct visible objects located at the surfaces of mirrors, echoes are not distinct sounds that occur at surfaces that reflect sound waves. Together, these four claims suggest that echoes are not distinct disturbance events that occur at reflecting surfaces.

The picture gestured at by analogy with mirrors is that hearing an echo is hearing the primary sound event over again. This is a re-encounter of a different sort from that rejected above. The sound event occurs only once, during a certain time interval. It can be perceived once, and then again during a later time interval because the waves it creates return. The sound neither travels nor returns to the perceiver; the perceiver experiences the same distal event over again because of the way the event's traces travel. Hearing an event that is past is thus like seeing an event that is past. Compare seeing a supernova from across the galaxy. If we could put mirrors in far outer space, we could see the same earthly event twice: once when it happened and once after its traces were reflected.

Why does the apparent distinctness of echoes from primary sounds, which I invoked against Nudds's argument, not tell equally against the claim that when the echo phenomenon occurs, we hear the very same event twice? We hear the event (qua echo) with distortion of location, but our experience as of the echo also occurs *later*. If an echo were an object experienced at a time later in its career, we would expect ordinary object-recognition to occur, given the echo's qualitative similarity to the sound initially heard. In fact, with objects we count on this sort of recognition to ground the perceived continuity of our material world. *Capgras Syndrome* is one form of *delusional misidentification syndrome* in which patients suddenly begin to believe that people and objects familiar to them have been replaced by exact qualitative duplicates; this failure of perceived continuity is notable and debilitating (see Breen *et al.* 2000). Events and time-taking particulars, however, are tied to a specific time and place when they occur. Though the 2002 World Cup Final might have been located at various times and places, it in fact occurred June 30, 2002, at International Stadium, Yokohama, Japan. That very event cannot occur again or elsewhere. Similar events experienced at different times and places are taken to be distinct events. So, if we happen to perceive the very same particular event over again, it should seem like a distinct event. Since echo phenomenology arises when we hear the very same sound event over again at a later time and different place, precisely what we should expect is the apparent distinctness of echo from primary sound.

On this model, we *can* perceive the same sound event twice because of how waves propagate. The situation is something like this: Suppose you hear the sound of the firework. You then travel faster than the sound waves, overtake them, and halt. You now hear the sound again—it seems to be in the same place it was before. We need not say the sound travels, only that the sound waves travel. Because of how information about sounds is transmitted through a medium, you are lucky enough to experience the same sound event over again. The medium disturbance you hear when you hear the sound for the second time is the very same disturbance event you heard earlier. Echo perception is similar. A reflecting surface, however, saves you the trouble of supersonic travel. You pay the price with distortion of location. The Event View nicely captures the correct way to conceive of echoes and echo perception.

9. Concluding Remarks

The Event View replaces the picture according to which sounds fill the air and travel as waves. Instead, sounds are events that occur where objects and bodies interact with the surrounding medium. Sounds are events that take place near their sources, not in the intervening space. Sound waves travel through the air carrying information about these distal events, and are the proximal causes of sound experiences in subjects; sound waves, however, are not sounds. The revision more accurately captures how we experience sounds to be.

The Event View is a natural account of what sounds are that avoids the dilemma concerning where sounds are located. It implies that sounds are distally located and stationary relative to their sources without making them solely the properties of material things. We should not accept the view that sounds are properties of objects themselves because we have good reason independent of the received view to think that sounds cannot exist in vacuums. The event that the Event theorist identifies as the sound cannot occur in the absence of a medium.

Taking sounds to be particular events of objects disturbing a surrounding medium furnishes a unified picture of what counts as a sound in cases that pose problems for any such theory. Sounds do not occur at barriers where transmission takes place. The phenomena accompanying constructive and destructive interference arise because of the spatial and temporal relations among sound sources and because information about sounds is transmitted by waves. Hearing an echo is hearing with distortions of place and time. The Event View entails no mysteries about sounds and sound experience.

References

Aristotle (1987). *De Anima*, in J. L. Ackrill (ed.), *A New Aristotle Reader*. Princeton: Princeton University Press.

Bennett, J. (1988). *Events and their Names*. Oxford: Clarendon Press.

Berkeley, G. (1975). 'Three Dialogues between Hylas and Philonous', in M. R. Ayers (ed.), *Philosophical Works*. London: Dent.

Blauert, J. (1997). *Spatial Hearing: The Psychophysics of Human Sound Localization*. Cambridge, Mass.: MIT Press.

Breen, N., Caine, D., Coltheart, M., Hendy, J., and Roberts, C. (2000). 'Toward an Understanding of Delusions of Misidentification: Four Case Studies', *Mind and Language*, 15: 74–110.

Gelfand, S. A. (1998). *Hearing: An Introduction to Psychological and Physiological Acoustics*, 3rd edn. New York: Marcel Dekker.

Kershaw, S. (2002). 'The Hunt for a Sniper: The Advice; Feeling that Witnesses Need a Hand, Police Offer One', *The New York Times*, 16 October.

Locke, J. (1975). *An Essay Concerning Human Understanding*, P. Nidditch (ed.). Oxford: Clarendon Press.

Nudds, M. (2001). 'Experiencing the Production of Sounds'. *European Journal of Philosophy*, 9: 210–29.

O'Callaghan, C. (2007). *Sounds: A Philosophical Theory*. Oxford: Oxford University Press.

Pasnau, R. (1999). 'What is Sound?' *Philosophical Quarterly*, 49: 309–24.

—— (2000). 'Sensible Qualities: The Case of Sound', *Journal of the History of Philosophy*, 38: 27–40.

3

Sounds as Secondary Objects and Pure Events

ROGER SCRUTON

1.

In this chapter I revisit the view, put forward in *The Aesthetics of Music* (Scruton 1997), that sounds exhibit the same ontological dependence on our perceptual experiences as secondary qualities, while not being qualities either of the objects that emit them or of the regions of space in which they are heard. Sounds, I suggest, are objects in their own right, bearers of properties, and identifiable separately both from the things that emit them and from the places where they are located. If you ask to what category of objects they belong, then I will say first that they are 'secondary objects', in the way that colors are secondary qualities (though what way is that?); second that they are events (though is there a relevant distinction between events and processes?); and third that they are 'pure events'—things that happen but don't happen *to* anything (though how is that possible?).

In ordinary physical events, such as crashes, physical objects undergo change: a car crash is something that happens to a car and to the people in it. But a sound is not a change in another thing, even if it is caused by such a change. Nor does anything participate in the sound in the way that the car participates in the crash. In my view, these distinctive features of sounds—that they are secondary objects and pure events—are fundamental to the art of music. They are also both curious in themselves and the source of interesting philosophical puzzles.

The theory of sounds as events has been defended by others, notably by Casati and Dokic (1994), and in more recent work by O'Callaghan (Chapter 2; see also 2007).[1] But those authors take a resolutely 'physicalist' approach,

---

[1] O'Callaghan (Chapter 2; 2007) argues that sound-events are 'object-involving', though they do not strictly take place in the object, consisting as they do in a disturbance that occurs at the interface of a source and a medium.

repudiating all suggestion that sounds might be essentially connected to the experience of hearing things, and identifying them instead with physical events in or around the objects that emit them. Sounds, they argue, are identical with neither the waves that transmit them nor the auditory experiences through which we perceive them. They are identical with the events that generate the sound waves—physical disturbances in physical things, such as those that occur when the string of a violin vibrates in air. In other words, sounds involve changes in the primary qualities of physical objects. According to this view, a sound happens to the thing that emits it, in the way that crashes happen to cars, and the happening consists in physical changes that could be measured in other ways besides hearing.

Physicalism about sounds is not confined to those who defend the view that sounds are events. Pasnau (1999, 2000), for example, who thinks of sounds as properties of the objects that emit them, is also a physicalist in my sense. He argues that an object 'has' or 'possesses' a sound when it vibrates at a particular frequency and amplitude; in other words, that sounds are primary qualities of the objects that are heard to emit them (in contrast to Locke, who describes sounds as secondary qualities in *Essay* II, viii). I do not explicitly argue against this form of physicalism, since I am persuaded that sounds are events, not properties.

2.

There are at least three motives for the physicalist approach. The first is scientific realism, which holds that all and only those things are real which are referred to (quantified over) in a true theory. We have auditory experience and make auditory discriminations, and can make judgments and predictions about the world on the basis of this. What explains this fact? Certain physical events, which act on the sense of hearing, by way of the medium of sound waves. What physical events? We answer that question with a primary-quality description of changes in the physical world that typically cause the emission of sound waves. If the explanation is complete, then so is the description. Sounds are the kinds of thing identified by such a primary-quality description. Any attempt to 'color in' the sound event in secondary-quality terms involves a detraction from and not an addition to its real physical essence. Of course sound events may also have secondary qualities, in the way that physical objects have colors. But these qualities are essentially tied to the perspective from which we perceive them, and are filtered out of the science that explains what we hear. Secondary qualities are mentioned but not referred to in the theory that explains them.

A second motive for the physicalist approach is evolutionary epistemology. What function, it might be asked, does the faculty of hearing serve? What information about the environment is made available by hearing, and how is hearing of special use to us in negotiating our passage through the world? To answer that hearing gives us information about sounds, when sounds are treated merely as the intentional objects of hearing, is to give no answer at all. (Useful information concerns material, not intentional, objects.) To answer that hearing gives us information about physical events that have a bearing on our survival is to make the first move towards an evolutionary explanation. And the next move follows immediately: sound waves bounce around obstacles and are difficult to block out. Hence hearing gives access to events that are too distant to touch and too hidden to see. We are beginning to understand why we have evolved the sense of hearing. Our theory seems to lead us naturally to the conclusion that sounds are events in the physical world, consisting in changes in the things that emit them—the gnashing teeth of the beast of prey, the energy of the approaching avalanche—things about which a bit of knowledge prior to a sighting comes in handy.

A third motive for the physicalist approach stems from the science of acoustics, and in particular from the theory offered by Helmholtz (1877/1954) of pitch, loudness, and timbre, which are the acoustic properties whereby we distinguish sounds and associate them with events in the physical world. We might be tempted to describe these as secondary qualities of sounds, or even as secondary qualities of sounding objects. However, as Helmholtz showed, we can map pitch onto frequency, loudness onto amplitude, and timbre onto the overtones of a sound wave. Frequency, loudness, and overtones are primary qualities of a physical object—the sound wave—and reflect similar qualities in the event (the vibration) that causes it. Those qualities of sounds that are not mapped by Helmholtz's sine-wave theory in a straightforward manner are nevertheless explicable in similar ways, as recording structural features of physical disturbances. All this adds weight to the suggestion that there is nothing more to a sound than the physical disturbance in which it originates, since every observable property of the sound can be correlated with, and explained by, some property of the disturbance.

3.

There is a more speculative reason for taking a physicalist approach to sounds, which is the difficulty of making sense of secondary qualities. The suggestion is commonly made that an object is red if and only if it is disposed to look red to normal observers in standard conditions. But this suggestion encounters a number of difficulties. Some argue that 'ungrounded dispositions' have no place in science and cannot form part of the ultimate reality. Hence it cannot be a brute fact that an object is disposed to look red. There must be something about it—the microphysical structure of its surface, for example—which explains the way it looks, and this explanation will identify what we are really referring to when we call something red (it will tell us the 'real essence' of redness). In which case the secondary quality vanishes, being at best part of the sense of 'red' but no part of its reference. Others argue that the definition of redness, in terms of 'looking red', is circular; still others argue that the definition of redness can be made non-circular only by supposing that we could give a purely phenomenological identification of the experience of 'looking red', an identification that made no reference to 'outer' things—a move that would immediately fall afoul of Wittgenstein's argument against the possibility of a private language. Others raise doubts about the terms 'normal' and 'standard', perhaps dismissing them as catch-all devices designed to give the appearance of an objective truth to what is merely a subjective impression.

All those difficulties are familiar from the literature, and it is well to make clear at the outset that I do not regard them as insuperable. The prejudice against ungrounded dispositions is no more than that. There is nothing incoherent in quantum mechanics, and quantum mechanics attributes ungrounded dispositions to the particles that it describes. The 'real essence' approach to secondary qualities is likewise an *ignoratio elenchi*, unless one says (with Locke) that in this case real essence and nominal essence coincide. No defenders of secondary qualities need deny that the disposition to look red to normal observers is grounded in some primary-quality (structural) property. But they will certainly deny that this property is identical with redness. For one thing, there may be more than one such structural property. For another thing, we don't have a concept of property identity that would enable us to decide this matter in favor of the skeptic.

As for the charge of circularity, I see it as providing the answer to the worry about the private language problem, rather than the ground for it. We can refer to an experience of red because we have a publicly applicable criterion for attributing that experience to another, namely the ability to discriminate red things by looking at them. The secondary quality, being publicly observable, gives us access to the inner experience of the one who perceives it. It is precisely because redness is a real property of objects that we can make sense of the idea that some things 'look red' to those who perceive them. I should add that I take the private language argument very seriously, as refuting the view that mental states have an inner 'essence' accessible only to the subject, and as refuting also the corollary to that view, according to which there could be an 'inverted spectrum'.[2]

Again, I see no difficulty in the invocation of the 'normal' observer and 'standard' conditions. For these are the presuppositions and the anchor of all our discourse. To invoke them in this context is not to introduce an element of subjectivity, nor even of 'perspectivalism'. There is nothing inconsistent in holding that redness is an objective property of whatever possesses it, and that redness is the disposition to look red to the normal observer. For we establish the normality of observers precisely by the way things look to them, discovering a deep-down harmony between their perceptions and ours, and observing that, by relying on their perceptions (including their perceptions of secondary qualities), they are able to predict and make use of their environment successfully, just as we do.

4.

I think we should be careful to distinguish perceptual capacities from emotional responses. In describing a joke as funny, I imply that the normal person would laugh at it, and perhaps this is what 'funny' means. However, 'normality' here is determined by cultural and not physiological norms. Maybe Japanese people would not laugh at the joke, despite their normality vis-à-vis the culture to which they belong. Moreover, laughter can be corrected, educated, criticized, brought (to some extent at least) within the province of the will, all of which means that the response is not wedded to the object but is part of the emotional and moral history of the subject. None of that is true of perceptions, which share with sensations their fixed physiological base. You cannot argue someone out of seeing red when that is what he sees, nor is the capacity to see red things

---

[2] See Scruton (1974) for the interpretation that I offer of Wittgenstein's argument. On the inverted spectrum hypothesis, see Block and Fodor (1972), as well as the entertaining rejection of the possibility by Dennett (1991).

as red one of the gifts of a culture. On the contrary, it is something we share with (some of) the animals. (This is one reason for believing that aspects are not secondary qualities, but at best 'tertiary' qualities of the objects that possess them. For aspects are subject to the will.)[3]

There are several different ways in which the perspective of the speaker can be invoked in giving the meaning of what he or she says. In an illuminating book, McGinn (1983) drives the comparison between secondary-quality descriptions and indexical sentences as far as it will go, in order to illustrate the ways in which there can be objective truths with an irreducibly self-referential component. 'I am over here' is not the kind of sentence that could play a role in a scientific theory. Nevertheless it can convey an objective truth, and sentences like this are unavoidable in recording the observations against which theories are tested. Secondary-quality ascriptions are also unavoidable in the testing of a theory, even if the aim of the theory is to explain them away.

It seems to me, therefore, that we should distinguish the cases where reference to the perspective of the subject undermines the objectivity of his or her judgment from cases where it is part of explaining it. Consider Kant's theory of 'transcendental idealism', according to which the perspective of 'possible experience' cannot be eliminated from our picture of the world. If Kant is right, then the world is the world as I know it, from my point of view. To accept this is not to accept that judgments about reality are either subjective or relative to a particular *person* or a particular *culture*. For the word 'I' in the Kantian theory stands proxy for 'any self-conscious being', 'any being who is able to identify the self in the first person', or 'any being who is able to ask whether the world is as it seems'. Only certain worlds contain self-conscious beings, and ours is one of them. But when we argue that our world is the world as perceived by and revealed to a self-conscious being, we open the way to no comparison. If Kant is right, then nothing said about our world stands open to correction from any superior point of view, since the corollary of his position is that there is no superior point of view—none, at least, that would be intelligible to us. Transcendental idealism shows how we might build our epistemic capacities into our account of reality, without reducing reality to a mere 'projection' of a perspective that we might not have had.

The case of secondary-quality descriptions is, of course, not exactly like that. Here we are describing the world, not as it is revealed to any self-conscious being, but as it is revealed to any being who shares our perceptual capacities. Hence the secondary qualities of things could change, as a result of a change

[3] See Scruton (1974).

in the faculties of the normal observer, and without any change in the primary qualities of the objects that he or she perceives. Such a change would be an objective change—a change in the way things are—and not just a change of perspective. For perspectives (at least as normally understood) are inherently open to comparison and can be altered, improved, educated, criticized, and abandoned. It is only when so understood that a perspective falls short of revealing the objective reality of the world on which it is turned. The case of secondary-quality perception resembles the self-conscious perspective studied by Kant, in that it does not involve such a 'falling short' of reality. The physiological basis of secondary-quality perception means that we have no rational access to any rival experience through which to correct it, so that the supposition that secondary qualities are merely 'perspectival' features of our world remains empty—a piece of pure philosophy, which merely spins around the definition of a secondary quality.

Since secondary qualities are rooted in our perceptual capacities, the science of secondary qualities is the science of perception. The matter to be explained by the theory of redness is our capacity to discriminate red things, not the fact of their being red. Concerning this second fact, science is silent, not because it denies the redness of things, but because redness denotes no explanatory property. You can explain the redness of things, but not by their being red. The defender of secondary qualities can readily accept that the explanation must be phrased in terms of primary qualities, and also can accept the possibility that we can discover a unique primary quality for each secondary quality (in this case the wave-band of refracted light).

The ability to discriminate (and therefore to classify) on the basis of secondary qualities is of enormous evolutionary advantage, given that by this means, food, predators, temperature, and so on can all be instantly and reliably assessed from a distance. The world comes color-coded and timbre-tried, and even creatures without eyes or ears must, like worms, evolve a capacity to navigate by the 'way things feel' to them; in other words, by secondary qualities of a tactile kind.

5.

In defending secondary qualities in that way, I may well be preaching to the converted;[4] and I have said nothing with which a 'physicalist' (someone

---

[4] Converted, for example, by McGinn (1983).

who thinks of sounds as physical disturbances, located in the objects that emit them) need disagree. The physicalist can admit that sounds have secondary qualities, and that it is by virtue of these qualities that we discriminate them. In themselves, however, so the physicalist will say, sounds are primary-quality events involving the bodies that emit them. Why should we deny that? Here are some of my reasons:

(i) I want to hold on to the view of sounds as 'objects of hearing' in something like the way that colors are objects of sight. Sounds are '*audibilia*', which is to say that their essence resides in 'the way they sound'. Hence they are absent from the world of the deaf person in the way that colors are absent from the world of the blind. The physicalist view has the consequence that deaf people could be fully acquainted with sounds (for instance by using a vibrometer which registers pitch and overtones), and also that people could see sounds without hearing them. There is nothing incoherent about either suggestion; however, both suggestions seem to relegate to the 'purely phenomenal' level everything in sounds that distinguishes them—not merely their relation to hearing, but (as I go on to argue) their internal order, their ability to speak to us, and much of their information-carrying potential.

(ii) I believe that we do not attribute the secondary qualities of sounds to the bodies that emit them, nor to events that occur in those bodies. We attribute them to the sounds themselves, conceived as independently existing events, located in a region of space. When I hear a car passing, what I hear is the *sound* of a car passing, an event caused by the car's passing but distinct from any event involving the car. The sound of the car is not an event in the car or a change in which the car participates. It is an event in itself.

(iii) The information conveyed by sounds is not, typically, information about a vibration in any object, nor do we usually group sounds together by reference to their source. Psychologists have studied auditory grouping, asking themselves what evolutionary function it might serve (Gelfand 1998). For example, we tend to group quiet sounds interrupted by bangs together, as though they formed a continuous sequence, just as we continue in imagination the lines on a page that are interrupted by blots. In general, sequences of sounds are 'streamed' in our perception—each allocated to a temporal *Gestalt*, formed according to temporal analogues of the principles for *Gestalt*-formation in vision. Proximity in pitch, duration, timbre, loudness, and so on lead to streams that endure through silences, interruptions, competing streams, and unstreamed events. Albert Bregman, perhaps the most noted researcher into these phenomena, is of the view that 'the perceptual stream-forming process has the job of grouping those acoustic features that are likely to have arisen from the same physical source' (1990: 138). However, as I go on to argue, streaming involves attributing to sounds an identity distinct from any process in their source, and involves the creation of a world of coherent sounds, rather than a world of coherent spatio-temporal objects. It is only *because* of this feature, indeed, that sound can fulfill its evolutionary function, that of providing an auditory map of the surrounding physical world.

(iv) Sounds can be detached completely from their source, as by radio or gramophone, and listened to in isolation. This experience—the 'acousmatic' experience of sound—removes nothing that is essential to the sound as an object of attention.[5] The striking thing is that sounds, thus emancipated from their causes, are experienced as independent but related objects, which form coherent complexes with boundaries and simultaneities, parts and wholes. There is a mereology of the acousmatic world that mimics the mereology of the world of physical objects, although it is dealing with temporal and not spatial wholes.

For those reasons, it seems to me that there is every reason to reject the physicalist view. In espousing physicalism, Casati and Dokic (1994: 179) bravely accept the consequence that sound is essentially non-phenomenal; in other words, that what a sound essentially is has nothing to do with how it sounds. They are prepared to accept that sounds have secondary qualities, but believe that the idea of a secondary object—an object all of whose properties are ways in which it appears—is 'extrêmement contestable'. But they do not say why, and their own account of sounds, which attributes to sounds only the most uninteresting of intrinsic properties, leaves us with no basis from which to explore *what we hear* through the organization of the auditory field. It seems to me that what we hear, both when we hear sounds in our day-to-day environment, and when we listen to sounds acousmatically, is not merely a subjective impression but a real part of the objective world. That is what I mean by describing sounds as secondary objects.

6.

Before substantiating those points, it is worth addressing the question of whether there are other examples of secondary objects. It seems to me that

---

[5] The use of the term 'acousmatic' in this context (borrowed from the Pythagoreans) is due to Schaeffer (1966), and has been taken up by Scruton (1997) and Chion (1999).

there are, and that they include two extremely important instances: rainbows and smells. I shall concentrate on rainbows, since they so closely parallel sounds. First, rainbows are *visibilia*, objects of sight that are not objects of any other sense. You cannot touch, smell, or hear a rainbow, and if when rainbows occurred, there were also sensations of touch (wetness in the air), of hearing (a fizzing sound), or of smell, we should feel no compulsion to say that we are feeling, hearing, or smelling the *rainbow*. (Lightning is always followed by a thunderclap, sometimes by a scorched smell, and often by rain, but we do not say that we hear, smell, or feel the lightning.)

Rainbows are secondary objects in the following sense: their existence, nature, and qualities are all determined by how things appear to the normal observer. That there is a rainbow visible over Sunday Hill Farm follows from the fact that a normal observer, located here, would have just such a visual experience if the observer looked toward the hill. This does not mean that rainbows have only secondary qualities: on the contrary, rainbows have many primary qualities, such as shape, size, and duration. But their having these qualities depends upon a counterfactual about experience.

Rainbows are located, but not precisely. There is no pot of gold at the end of the rainbow because there is no place that is the end of the rainbow; nor is there a stretch of sky that the rainbow occupies. In this case, location, too, is experience-dependent. To say that there is a rainbow visible over the hill is to say that a person located in a certain place and looking toward the hill would see the arch of a rainbow lying over it. Rainbows don't take up space, and don't exclude other objects from the spaces where they appear. Nevertheless, there is a distinction between the places where rainbows are and the places where they are not.

Rainbows are real and objective. Someone who claims to see a rainbow where the normal observer could not see one either is under an illusion or has made a mistake. This person is wrong about the way the world is. Rainbows can be other than they seem, and seem other than they are. There were rainbows in the world before there were creatures to observe them, for the truth about rainbows consists in the truth of a counterfactual, concerning what the normal observer *would* see were the observer's eyes to be turned in a certain direction. The rainbow, like the photon, is an ungrounded disposition, and it illustrates the way in which ungrounded dispositions can be part of the fabric of reality.

None of that implies that we could not give a full explanation of rainbows in terms of the primary-quality structure and changes of normal primary objects. The explanation is indeed familiar to us, and invokes light waves and their refraction by water droplets in the air. This explanation of rainbows is very similar to the explanation of the experience of sound, in terms of the transmission of a vibration by sound waves. But it will not mention any particular object that is *identical* with the rainbow, in the way that the physicalist urges us to consider the vibration of the source as *identical* with the sound. Hence it will leave us free to locate the rainbow at the place where it appears, and not at some place chosen for its prominent role in the theory of rainbows (for instance the patch of water drops in the air, the sun, the eye of the beholder).

7.

If we ponder the example of rainbows—and the equally illuminating example of smells—we will see, I believe, that the physicalist theory of sounds is unmotivated. There is every reason to treat sounds as *audibilia*, in just the way that we treat rainbows as *visibilia*. In doing so, we deny nothing that the physicalist affirms, other than a statement of identity that renders accidental all those features of sounds that we are normally disposed to treat as essential—features, in short, of *the way they sound*. By treating sounds as secondary objects, we restore to them their true nature, as information-bearing events that are organized aurally.

Sounds are not 'reidentifiable particulars' of the kind singled out by Strawson (1959) to be the anchors of our conceptual scheme. Nor are they properties of the things that emit them. They exist as both tokens and types—'the sound of a steam train', for example, may refer either to a token sound or to a sound type, depending on context. Sounds are located in space but have no spatial boundaries. They take time, and usually have a beginning and an end in time. However, their individuation and identity are contentious. We do not have a single, hard, and fast criterion for determining when one sound ends and another begins, or when a complex is composed of two simpler sounds rather than one complex sound, or when a sound heard now is numerically the same as a sound heard then.

None of that is surprising if sounds are events. For events have fluid conditions of identity, take time but are not necessarily reidentified across time, can be simultaneous, composite, and only vaguely located, all in the manner shown by sounds. This does not, of course, get us very far toward understanding what sounds are, since events are notoriously problematic from the metaphysical point of view. We know that we need them in our ontology; but we don't know how to count them, nor how to assign conditions of identity to them, various brave attempts notwithstanding.[6] I am inclined to the view that procedures for counting, individuating, and reidentifying events are profoundly interest-related, and have little or no bearing on the ultimate structure of reality. Events would be true individuals, with fixed conditions of identity, only if time stood still. But time, like an ever rolling stream, bears all its sons away.

Nevertheless, to say that sounds are events is not to end the story in a familiar mystery. For there is another and less familiar mystery that must now be broached, the mystery of the pure event. Those influenced by Aristotelian conceptions of substance, or Strawsonian conceptions of ontological dependence, are apt to identify events in terms of the changes undergone by particulars. A car crash consists in a car's crashing, and all the changes to all the other objects that are affected. Fully to describe the event is to identify all the objects that are changed together with their changes. Most descriptions of events, on that view, are partial descriptions, which single out some salient object and describe the changes undergone by *it*. And all descriptions of events proceed by identifying the objects that participate in them.

The reasons for taking that view are implicit in the first chapter of *Individuals* (Strawson 1959) and also in Wiggins's (2001) more Aristotelian approach. Our conceptual scheme, it is suggested, is anchored in acts of identification, both of individuals and of kinds. And both individuals and kinds stand in relations of 'ontological priority' to each other. *A* is ontologically prior to *B* when identifying (a case of) *B* involves identifying (a case of) *A*, but not vice versa. (Thus, particulars are ontologically prior to their non-essential properties.) Events can be fitted into our ontology if we think of them as transformations undergone by particulars: they involve no additional act of identification beyond that of identifying an individual or substance and describing its properties over time. Individuals, on this view, are ontologically prior to events, and the suggestion that an event might be identified without identifying an individual that undergoes change is ruled out by the underlying presuppositions of public reference.

Once we acknowledge the existence of events that are also secondary objects, however, the way is open to the 'pure event': the event that happens, even though it does not happen *to* anything (except perhaps to itself), the event that cannot be reduced to changes undergone by reidentifiable particulars. This, I believe, is how we should think of sounds. They occur, but they stand alone,

---

[6] See, for example, Kim (1976); Bennett (1988), which contains an effective criticism of Kim; and Davidson (1980).

and can be identified without identifying any individual that emits them. Of course, our language for characterizing sounds tends to describe them in terms of their normal source—dripping, croaking, creaking, barking. But reference to a source is not essential to the identification of the sound, even when it is compelled by the attempt to describe it. It is in some sense an accident if we can attribute a sound to a particular—to say that it is the sound *of* this thing, caused by changes *in* that thing, and so on. It would be quite possible for us to be surrounded by sounds, like Ferdinand on Prospero's Isle, which we individuate, order, and interpret without assigning to any of them a physical process as origin or cause.[7]

This gives me another reason to be dissatisfied with the physicalist view. For it seems to tie sounds too firmly to their sources. In particular, it seems to suggest that our ordinary ways of identifying sounds, as self-dependent events that bear their nature in themselves, is mistaken. The physicalist view banishes to the margin those features of sound that make sound so important to us, not only epistemologically, but also socially, morally, and aesthetically. In particular, it does not recognize the 'pure event' as a distinct ontological category, and one that introduces unique possibilities of communication.

8.

It is here that we should return to the psychologist's investigations into auditory streaming. Our ears are presented at every moment with an enormous perceptual task, which is to group sounds together in such a way as to make sense of them—either by assigning them to their causes, or by discovering what they mean. This 'auditory scene analysis', as Bregman (1990) describes it, is carried on continuously, and operates by amalgamating auditory episodes into temporal *Gestalten*. In certain applications, this search for the good *Gestalt* can be explained as the evolutionary epistemologist would explain it, namely as a first step towards tracing a sound to its cause. However, it differs in a crucial respect from the search for the good *visual Gestalt*.

Suppose you are looking at a dot-picture, and unable to make out the figure that the dots compose. And suppose that suddenly the figure 'dawns' on you, and there, before you, in the unseen lines that join the dots together, is the outline of a face. The joined up *Gestalt* is unified by a shape, and the shape is one that you recognize. It is quite clear how this ability to amalgamate

---

[7] See the example of the 'Music Room' discussed in Scruton (1997: ch. 1).

bits of visual information into a whole assists us in recognizing objects. For the perceptual *Gestalt* shares properties such as shape, size, and color, with its object, and the emergence of a visual *Gestalt* involves conceptualizing visual experience in terms that the object, too, fits. In describing the order of the *Gestalt*, you unavoidably refer to an order of objects.

Auditory scene analysis is quite unlike that. Streams are not described or recognized through properties that could be exhibited by the physical events that produce them. Sounds that succeed each other on the pitch continuum seem to 'flow' along the continuum; tones an octave apart are heard as parts of a single tone (as when a congregation sings a hymn in what sounds like unison). Dynamic properties of the sound such as attack and crescendo lead to streamings that seem entirely internal to the world of sound, bearing no relation to the real sequences that produce them. It seems that we have an inherent tendency to group sound events as 'auditory figures' without making bridges to the physical world.[8]

None of that should surprise us, if sounds are pure events, understood in isolation from the objects that emit them, and grouped according to their intrinsic properties. We can see what this means by reflecting on the scale illusion discussed by Deutsch (1999), illustrated in Figure 3.1. Headphones are placed over the subject's ears, and the notes of a descending scale and an

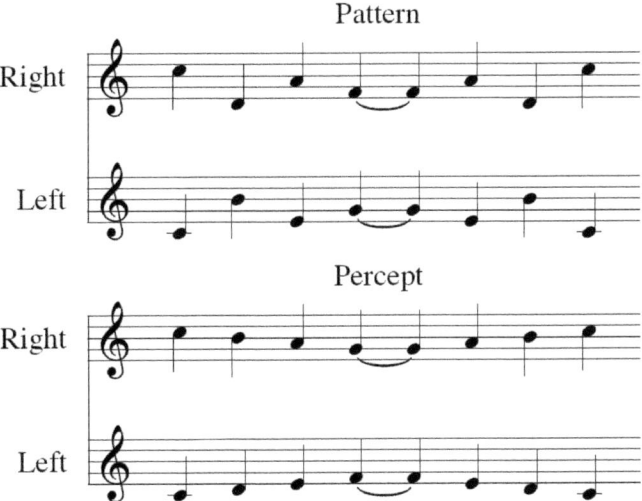

Figure 3.1. The scale illusion
Source: Reprinted with permission from Deutsch (1999: 322).

---

[8] See Divenyi and Hirsh (1978) and Bregman (1990: 141).

ascending scale played in each ear. The two ears receive, respectively, the inputs indicated by the pattern at the top of Figure 3.1. What they hear, however, is indicated by the percept at the bottom of Figure 3.1. In other words, the sounds are arranged by their own internal gravity as two cogent sequences, even though those sequences are played into neither ear, and represent no causally unified process in the physical world. Streaming is useful to us because, in the normal run of things, the streams that we hear correspond to such causally unified processes. But this correspondence is not *what* we hear, even when it *explains* what we hear.

9.

The ability of pure events to stand in perceived relations to each other independent of any perceived relations between their causes is a deep presupposition of music, in which note follows note according to the internal logic of the musical line, giving rise to a virtual causality that has nothing to do with the process whereby the sounds are produced.[9] That is not to say that the kind of 'streaming' that goes on in musical hearing is the same as the streaming of ordinary sound perception: in my view it is not, since it is shaped by spatial metaphors that are the product of a musical imagination (Scruton 2004: 184–7). Nevertheless, music is an extreme case of something that we witness throughout the sound world, which is the internal organization of sounds as pure events, 'detached', as Samuel Beckett puts it, 'from the sanity of a cause' (1965: 23).

That is not to deny that auditory scene analysis is connected to the attempt to gather information about the physical events that cause us to hear sounds. It is rather to insist that sounds cannot be related easily to their causes by the perceiving ear, and are not imprinted, in the manner of visual images, with either the contours or the location of the things that produce them. If it is to make sense of the chaotic 'manifold' of impacting sound events, therefore, the ear must order them in ways that permit us to distinguish continuous processes from isolated events, simultaneous from complex sounds, and so on, without relying upon any observation of physical objects and their changes. That is precisely what is made possible, if sounds are pure events. For pure events contain within themselves the principles whereby they can be ordered—principles of aggregation and disaggregation whereby events can

---

[9] Consider Brahms's second piano concerto as discussed by Scruton (1997: ch. 2) and the further argument in that chapter concerning the general distinction between sound and tone.

be decomposed into smaller events, joined up into larger ones, and accorded precise relations in time, all without stepping into the order of things.

10.

There is another matter that needs to be touched on in this connection, which is the effect of the human voice on hearing. Aristotle makes the striking suggestion that the voice is distinct from all other objects of hearing since we hear it in another way (1987: 420b). First, he defines voice (*phone*) as the kind of sound that is made by a creature with a *psuche*—in other words, the *sound of the soul*. Voice is also a 'kind of sound with meaning' (*semantikos ... tis psophos estin he phone*)—voice is the *sound of meaning*. Aristotle's suggestive remarks gain some support from the following features of vocal sounds: (i) Unlike ambient sounds, they are *addressed* to, and not merely heard by, others. (ii) They originate in, reveal, and express a center of consciousness—'a way the world seems'. (iii) They are, or include, language, and therefore exhibit grammatical order. (iv) Hence, they convey information that is not (typically) information about their cause, but information about the world, encoded in their own symbolic structure.

All those features of vocal sounds make use, I believe, of their nature as secondary objects and pure events. I order sounds according to intrinsic relations, which allows me to form longer sequences out of shorter, and to form separate streams out of a continuous babble; thus, I can hear sounds as addressed to me, ordered grammatically, and conveying a message. In hearing your voice, I hear this kind of intrinsic sonic order, and attribute it not to the vocal chords or even to the body that contains them, but to you, as another subject. In an important way my grasp of your utterance bypasses the search for causes, on its way to a meaning that I can attribute directly to you. That perhaps is what Aristotle had in mind in saying that I hear the voice in another way: I hear it as detached from the natural order, so to speak, existing in a realm of pure communication. To put it in a more Hegelian idiom, speech sounds have been lifted from the world of objects into the world of subjects, and are intrinsically addressed from me to you or from you to me. (Cf. Homer's metaphor of 'winged words'.) Hence the voice in the cinema calls the eye toward the person to whom it can be attached, or, if there is no such person in view, to the perspective of an unseen observer.[10] The disembodied voice

---

[10] See the illuminating discussion in Chion (1999).

can *haunt* the cinematic image, so that it ceases to be a neutral scene awaiting action and becomes 'seen by the unseen'.

Our way of hearing voices impacts elsewhere on our auditory experience. The experience of grammar inhabits our ears, so to speak, leading us to hear grammatical or quasi-grammatical order in all kinds of sequences that do not in fact exhibit it. The experience is particularly powerful in music, which often gives the impression of being addressed to us in something like the way that speech is addressed, so that we turn to it with ears already attuned to meaning. This experience is an important motive for the view that music communicates states of mind by virtue of a quasi-semantic structure.

11.

This brings me to the final reason for being dissatisfied with the physicalist theory of sounds, which is music. It seems to me that a theory of sounds ought to make sense not only of ordinary hearing, but also of all those special acts of attention of which sounds are the object. We hear sounds, just as animals hear them. We also listen to them, listen out for them, attend to them, and so on. All those mental acts could be accounted for on the physicalist assumption, as acts directed toward physical vibrations. In the case of music, however, we hear an order, which, while intrinsically auditory, is unperceivable to the animals, and dependent on our ability to detach sounds entirely from their physical cause. This is the order granted by the acousmatic experience. Although this experience grows from and exploits the grouping, streaming, and aggregating principles of ordinary hearing, it is also, I argue, shaped by an intentionality of its own. In conclusion, I shall illustrate the way in which the peculiar metaphysical status of sounds endows them with the plasticity required by the musical experience.[11]

The intentionality of perceptual experiences is determined by the concepts that inform them—either through perceptual beliefs or through acts of imagination. When listening to music we attend to sequences, simultaneities, and complexes. But we hear distance, movement, space, closure. Those spatial concepts do not literally apply to the sounds that we hear. Rather, they describe what we hear *in* sequential sounds, when we hear them as music. In other words, the concepts that provide the fundamental framework for musical perception are applied metaphorically, in the act of acousmatic attention.

[11] See Scruton (1997) for a more detailed discussion.

In making sense of this imaginative transfer of concepts from the world of physical space to the world of tones, we are obliged, I believe, to treat sounds as I have treated them: as secondary objects that are also pure events. The sounds themselves form the ground in which we hear musical movement, in something like the way that the colored patches on a canvas form the ground in which we see the face. This experience of movement is possible, however, only because sounds are heard to occupy places on the pitch spectrum. They map out distances and points on a one-dimensional continuum. They form recognizable auditory types, classified by timbre, pitch, and duration. They define the boundaries of musical space, places where melodies begin and end, and so on.

The physicalist might say that, in describing those features of sounds, I have merely described secondary qualities of physical vibrations. However, I believe that it is only because sounds are objects, all of whose properties lie in the realm of the audible, that they can be organized as tones. When so organized, they serve to pin down the auditory space in something like the way that physical objects pin down physical space—occupying places, moving between places, and standing in spatial relations.

In hearing sounds as music, we also hear them as pure events that we can reidentify—as when we recognize a melody sounding in another key, with another timbre, or at another pitch. Because we detach sounds in imagination from their physical causes, they become individuals for us, with a life of their own. They can amalgamate with other pure events to produce larger events; they can divide and fragment; they can occur in the background or in the foreground, on top or underneath, augmented or diminished, half hidden, ornamented, varied, and implied. Hearing sounds in this way, we rely at every point upon our ability to deal with them as real individuals, while neither reflecting on nor hypothesizing the background causality from which they arise.

References

Aristotle. (1987). *De Anima*, In J. L. Ackrill (ed.), *A New Aristotle Reader*. Princeton: Princeton University Press.
Beckett, S. (1965). *Proust: And Three Dialogues with Georges Duthuit*. London: J. Calder.
Bennett, J. (1988). *Events and Their Names*. Oxford: Clarendon Press.
Block, N. and Fodor, J. (1972). 'What Psychological States are Not'. *Philosophical Review*, 81: 159–81.
Bregman, A. S. (1990). *Auditory Scene Analysis*. Cambridge, Mass.: MIT Press.

Casati, R. and Dokic, J. (1994). *La Philosophie du Son.* Nîmes: Chambon.
Chion, M. (1999). *The Voice in Cinema,* C. Gorbman (translator). New York: Columbia University Press.
Davidson, D. (1980). 'The Individuation of Events', in *Essays on Actions and Events.* Oxford: Clarendon Press.
Dennett, D. C. (1991). *Consciousness Explained.* London: Little, Brown.
Deutsch, D. (1999). 'Grouping Mechanisms in Music', in D. Deutsch (ed.), *The Psychology of Music,* 2nd edn. New York: Academic Press, 299–348.
Divenyi, P. L. and Hirsh, I. J. (1978). 'Some Figural Properties of Auditory Patterns'. *Journal of the Acoustical Society of America,* 64: 1369–86.
Gelfand, S. A. (1998). *Hearing: An Introduction to Psychological and Physiological Acoustics,* 3rd edn. New York: Dekker.
Helmholtz, H. (1877/1954). *On the Sensations of Tone,* 4th edn. New York: Dover.
Kim, J. (1976). 'Events as Property Exemplifications'. in M. Brand and D. Walton (eds.), *Action Theory.* Dordrecht: Reidel, 159–77.
McGinn, C. (1983). *The Subjective View: Secondary Qualities and Indexical Thoughts.* Oxford: Oxford University Press.
O'Callaghan, C. (2007). *Sounds: A Philosophical Theory.* Oxford: Oxford University Press.
Pasnau, R. (1999). 'What is Sound?' *Philosophical Quarterly,* 49: 309–24.
—— (2000). 'Sensible Qualities: The Case of Sound'. *Journal of the History of Philosophy,* 38: 27–40.
Schaeffer, P. (1966). *Traité des Objets Musicaux.* Paris: Seuil.
Scruton, R. (1974). *Art and Imagination: A Study in the Philosophy of Mind.* London: Methuen.
—— (1997). *The Aesthetics of Music.* Oxford: Clarendon Press.
—— (2004). 'Musical Movement: A Reply to Budd'. *British Journal of Aesthetics,* 44: 184–7.
Strawson, P. (1959). *Individuals.* London: Methuen.
Wiggins, D. (2001). *Sameness and Substance Renewed.* Cambridge: Cambridge University Press.

4

Sounds and Space

MATTHEW NUDDS

Where are sounds and where do we experience them to be? In what follows, I argue that when we hear sounds, we hear them where we are. We do not hear them *to be* where we are, however, because we do not hear sounds to be anywhere. This conclusion follows from the account I give of what sounds are (Section 1), and of the role of space in auditory perception (Section 2).

1. What are Sounds?

The Function of Auditory Perception

We have perceptual systems because they are useful, and they are useful because they tell us about our environment and about what is happening in our environment. Auditory perception is no exception: it tells us about objects and events in our environment by detecting disturbances in the air caused by those objects and events.

In what follows, I am going to talk about the perception of what may be labeled 'ecological' sounds and their sources—those sounds produced by naturally occurring events of various kinds. I am going to ignore the role of sounds in communication and in the perception of music. Although both depend on the more fundamental processes involved in ecological sound perception that I am going to describe (and to the extent that they do not depend on those more fundamental processes they are distinct perceptual systems), they raise specific problems that I will not go into here.

Auditory perception tells us about objects and events in our environment and the sounds produced by those objects and events. When we hear a sound we can attend to the source of the sound and its properties—what it is, how it is, where it is, whether it is moving, and so on; or, we can attend to the sound itself and its properties—its pitch, timbre, loudness, and so on. Although relatively little investigation has been done to determine how good we are at perceiving and recognizing sound sources, what has been done has found that we are surprisingly good at both.

For example, we can perceive the size of an object dropped into water, that something is rolling, the material composition of an object from the sound of an impact, and the force of an impact. We can distinguish objects with different shapes, and we can tell the length of a rod dropped onto the floor. When something is dropped, we can hear whether it bounced or broke. We are good at recognizing natural sounds, such as footsteps, hands clapping, paper tearing, and so on. We can tell that a cup is being filled and when it is full. These are all examples that have been empirically demonstrated; it would be possible to find many more.[1]

Ecological sounds themselves, rather than their sources, are not very interesting. For many ecological sounds, it is difficult to attend to the sound rather than to the source of the sound, and we are poor at describing the properties of the sound as opposed to the source of the sound—we tend to describe the sounds in terms of their sources, as in 'it's like the sound made by ...' (Gaver 1993a). It is easy to overlook this about our perception of ecological sounds because we have become so used to hearing artificially produced sounds for which it is not true (or for which the opposite is true).

We can perceive these various things about the sources of sounds because auditory perception functions to tell us about the sources of the sounds we hear, and our auditory experience is of, or represents, the sources of sounds and their properties as well as sounds and their properties. It does so, I shall argue, by representing sounds as having been produced by their sources.

The Sources of Sounds

To understand how auditory perception carries out its function of telling us about the sources of sounds, we need to start with how the things that produce sounds—the sources of sounds—do so. Sounds can be produced by different kinds of things—liquids, material objects, strings, air movement—and in different ways. For the sake of simplicity, I am going to consider only material objects. Material objects produce sounds when they are struck, tapped, scraped, broken, or otherwise caused to vibrate.

We generally picture vibrations as simple sine waves, but not even something as simple as a plucked string vibrates in a simple way. The vibration of

---

[1] See, for example, Cabe and Pittenger (2000), Carello *et al.* (1998, 2005), Freed (1990), Kunkler-Peck and Turvey (2000), Schiff and Oldak (1990), VanDerveer (1979), Warren and Verbrugge (1984), Wildes and Richards (1988).

a plucked string is complex and comprises a number of simple vibrations at different frequencies that are integer multiples of the lowest, or fundamental, frequency of the vibration.[2] Objects vibrate along a greater number of dimensions than strings, and consequently their vibrations have a greater number of frequency components. Any complex vibration is equivalent to a number of simple sine wave frequency components superimposed on each other with appropriate phase relationships. That means we can represent any complex vibration as a pattern or structure of individual phase-related frequency components.

What's important for our understanding of auditory perception is that the particular pattern of frequency components produced by a material object when it vibrates is determined in a law-like way by the physical nature of the object and the nature of the event that caused it to vibrate.

For example, the shape and size of the object determine the lowest frequency of its vibration, and what harmonics are present. The overall amplitude of the vibration is determined by the force that initially deforms the object. Because objects are not linearly elastic, the amplitude of individual frequency components varies with the initial deformation. The spectral composition of the vibration therefore changes according to how hard the object was struck (which is why we can distinguish in experience the intensity or loudness of a sound from the force of the impact that produced the sound; the first is a property of the sound, the second of the source of the sound). Vibrating objects lose energy over time, and their vibration decays. The rate of decay of different frequency components is determined by the material of which the object is composed.

Because the pattern of frequency components that comprise the vibration of an object and the way that pattern changes over time is determined by the nature of the object and the events that caused it to vibrate, that pattern and the way it changes provide a great deal of information about the object and the interaction that produced the vibration.

The vibrations of objects are transmitted through the air. In an enclosed space, the vibrations will tend to reflect off hard surfaces and surrounding objects, and vibrations from different objects will interact with each other. These reflections and interactions will change the spectral composition of a vibration in determinate ways. At any place, the local disturbance of the air at that place will carry information about any number of events and,

---

[2] The vibration of a plucked string is made up of the odd harmonics of the fundamental; unlike the vibration of a string excited in some other way, which includes both odd and even harmonics. See Fletcher and Rossing (1998: ch. 2).

in virtue having been structured by it, about the environment in which those events occur. This local disturbance of the air is what is detected by our ears.

From Detection to Perception

Auditory perception tells us about the sources of sounds. In order to do so, it must extract the information about those sources embodied in the pattern of frequency components of the sound wave that is detected by the ears. The frequency components detected by the ears are the result of the interaction of many different object vibrations. How does the auditory system extract information about individual objects?[3] We can divide the process into three stages.

The first stage is that of sensory transduction or detection: the ears detect properties of the sound wave—the local disturbance of the air. We can think of the result of the detection of the sound wave as a temporal spectrogram of the sound wave which encodes the frequency and temporal properties of the sound wave's vibration. In effect, the ears detect each of the frequency components (within a detectable range) present in the sound wave's vibration.

Information about objects and events is embodied in the relationships among the frequency components produced by an object's vibration; but the frequency components detected by the ears may have been produced by many different sources; so, in order both to determine how many sound-producing events are occurring at any time and to extract information about the objects involved in them, the auditory system must organize frequency components into groups corresponding to the objects and events that produced them. The second stage of processing therefore involves grouping together frequency components that

---

[3] Albert Bregman (1990: 5–6) illustrates the problem as follows. Suppose you are standing by a lake on which there are boats:

Your friend digs two narrow channels up from the side of the lake. Each is a few feet long and a few inches wide and they are spaced a few feet apart. Halfway up each one, your friend stretches a handkerchief and fastens it to the side of the channel. As waves reach the side of the lake they travel up the channels and cause the two handkerchiefs to go into motion. You are allowed to look only at the handkerchiefs and from their motions to answer a series of questions: How many boats are there on the lake and where are they? Which is the most powerful one? Which is the closer? Is the wind blowing? Has any large object been dropped suddenly into the lake? Solving this problem seems impossible, but it is a strict analogy to the problem faced by our auditory systems. The lake represents the lake of air that surrounds us. The two channels are our two ear canals, and the handkerchiefs are our ear drums. The only information that the auditory system has available to it, or ever will have, is the vibrations of these two ear drums. Yet it seems able to answer questions very like the ones that were asked by the side of the lake: How many people are talking? Which one is louder, or closer? Is there a machine humming in the background?

have been produced by the same source. Frequency components need to be grouped so that all the frequency components produced by a single source are treated together, and those from different sources treated as distinct by subsequent processes.

There are two kinds of grouping. Frequency components produced at a time must be grouped together as having been produced *simultaneously* by a source; simultaneous groups must be *sequentially* grouped over time as having been produced by a single temporally extended event, and series of such sequences grouped as having been produced by a series of events involving the same object.

For example, when two objects make a sound simultaneously, we normally hear two distinct sounds—we hear the sound made by each of them. When we hear water filling a glass, we hear a single continuous sound—we experience earlier and later parts of the sound as parts of the same sound; when we hear an object dropped onto a hard surface and bounce, we hear the sound of each individual bounce and we hear the series of bounces as the bounces of a single object. We hear these sounds as a consequence of the way that frequency components are grouped. This grouping is necessary in order for subsequent auditory processes to extract the information about sources embodied in the sound wave.

How does the auditory system group frequency components, and how does it determine which frequency components to group together and which to group separately? There are relationships that exist between components produced by the same source that are unlikely to exist between components produced by different sources. For example, an object's vibration often has frequency components that are harmonics of a fundamental frequency, and so the frequency components of a sound wave that are produced by the same source will often be harmonically related. Such harmonic relationships are unlikely to exist between frequency components produced by distinct sources, since it is unlikely that two simultaneously occurring natural events produce overlapping sets of harmonics. This means that if the auditory system detects a number of frequency components that are harmonically related, then they are likely to have been produced by the same source. Similarly, the sound wave produced by a single event will have frequency components that share temporal properties—all the components will begin at the same time, are likely to be in phase with one another, and are likely to change over time in both amplitude and frequency in similar ways. Components produced by distinct sources are very unlikely to be related to each other in these ways. This means that if the auditory system detects a number of frequency components that share temporal properties, then they are likely to have been produced by the same source.

When the auditory system detects these relationships between components, it groups them together and treats them as having been produced by the same source. Components that are not related in this way are grouped separately. Similar relationships exist between frequency components produced by the same source at different times.

These examples are of properties that determine bottom-up or stimulus-driven grouping. It is likely that the auditory system also uses information in a top-down way to determine grouping, particularly of sequences. Some sequences are grouped because they fit into a pattern that the auditory system recognizes as likely to have been produced by a certain kind of source. Hearing an object as bouncing, for example, may be the result of top-down grouping.[4]

It is important to note that we cannot explain why the auditory system groups the frequency components that it detects in the way it does, other than in terms of a process that functions to extract information about the objects that produced those frequency components. This is true of both simultaneous and sequential grouping. The auditory system groups together all and only frequency components that are likely to have been produced by the same source *because* they are likely to have been produced by the same source.

The third stage of processing is not well understood. The process of grouping results in sets of frequency components that are treated by subsequent processes as having been produced by a single source. These sets of components carry information about those sources, and the fact that we can perceive various properties of the sources of sounds means that the auditory system must extract that information. Exactly what information is extracted and how it is extracted, is, for the most part, unclear. We can perceive how many sources there are and often where they are. I have described examples of various features of sources that we can perceive, and examples of our ability to recognize sources as events of certain kinds or as involving certain kinds of object. These recognition processes might match representations of the features of sources with representations of kinds of events and objects (similar to the way visual object recognition functions), or they might simply track some characteristic pattern of frequency components produced by certain kinds of events and objects. However exactly the information extraction and object recognition processes work, they must be sufficient to explain our capacity to perceive and recognize the sources of sounds.

---

[4] Bregman (1990: 397) calls this 'schema-based organization'. It involves 'the activation of stored knowledge of familiar patterns or schemas in the acoustic environment'.

In summary, I have characterized auditory perception as having the following functional organization:

Source events → *sound wave* → *sensory transduction* → *grouping (simultaneous and sequential)* → *extraction of primitive features (←)* → *recognition (←)*

Auditory Experience

It is in virtue of the operation of these psychological processes that we have experiences of sounds and their sources. What sounds we experience and how we experience them to be is determined by the way the auditory system groups the frequency components it detects: the sounds we hear correspond to frequency component groupings. If the auditory system groups the components it detects into a single group, then we experience a single sound; if it groups them into two groups, then we experience two sounds. Given that the auditory system groups frequency components that are likely to have been produced by the same source, the sounds we experience normally correspond to their sources—to the things that produced them.

Given that what sounds we hear is determined by the way the auditory system groups frequency components, and that we can only explain why the auditory system groups frequency components in terms of a process that functions to tell us about the sources of sounds, it follows that we can only explain why we experience the sounds we do in terms of a process that functions to tell us about the sources of sounds.[5]

What are Sounds?

The auditory system functions to tell us about the sources of sounds. We experience sounds as a consequence of the way the auditory system carries out this function: the sounds we experience are the result of the way the auditory system groups the frequency components that it detects in order to extract information about the sources that produced them.

Given this, it is plausible that our experience of sounds represents patterns or structures of frequency components instantiated by the sound waves that are detected by the ears. It follows that an experience of a sound is veridical just in case it is produced by the pattern or structure of frequency components that would normally produce that experience. It is not veridical if it is not

---

[5] This reverses the order of explanation assumed by most, if not all, other accounts of sounds. There is not space here to defend my account in detail; for a more detailed discussion see my (unpublished a).

produced by any such pattern, or if it is produced by a pattern that would not normally produce that experience.[6]

Our normal ways of individuating sounds allow that two people in different—even very distant—places can hear the same particular sound. You and I both hear the same sound when we hear the sound of a gunshot. To deny this would be to allow that a single event—a gunshot, say—produces more than one sound: a sound heard by me, and a sound heard by you at a distance. Since people at different places who hear the same sound are not, or need not be, affected by the same instantiation of frequency components, we cannot identify particular sounds with *instances* of a pattern or structural type. Similarly, our normal ways of individuating sounds allow that two people hear the same sound even if they hear it as having different qualities. The sound of a gunshot heard close by may be different—louder, sharper—from that same sound heard at a distance; it is, nonetheless, the same sound. But again, since the instantiation of frequency components must be different in the two cases, and may even be an instance of a different pattern, the sound we both hear cannot be identical to an instance of a pattern type.

If sounds are not identical to instances of pattern types, then could they be the pattern types themselves? Our normal way of individuating sounds treats them as particular things such that we can allow that two sounds may be qualitatively the same—the same type of sound—and yet be distinct, individual sounds. Two sounds that are indistinguishable, for example, are usually counted as distinct if they are produced by different sources. Thinking of sounds as pattern types does not allow us to make this distinction. Furthermore, if sounds are things that come into being when they are produced, then for any sound there is a time before which it did not exist, a time at which it came into existence, and, presumably, a time at which it will cease to exist. Although instances of pattern types have these temporal properties, pattern types themselves do not.

Any account of sounds should, as far as possible, accommodate our normal ways of individuating sounds. The ontological category that comes closest to doing so is that of particularized types or abstract individuals. To view sounds as abstract individuals would be to view them as belonging to the same ontological category as symphonies and other multiply instantiated art works or words (on Kaplan's account of the ontology of words).[7] To claim that sounds

---

[6] For further discussion, see my (unpublished a).

[7] This idea that sounds are abstract individuals was suggested to me by Mike Martin. For Kaplan's account of the ontology of words, see Kaplan (1990).

are abstract individuals is not, of course, to deny that sounds are instantiated by sound waves any more than to claim that words are abstract individuals is to deny that words are instantiated by, for example, patterns of ink on paper. It simply allows the possibility that a sound, like a word, may be instantiated at more than one place and time.

Where are Sounds?

Having given an account of what sounds are, I am now in a position to address the question of where they are located. I have suggested that sounds are patterns or structures of frequency components instantiated by sound waves. If that is right, then we can only make sense of talking of the location of a sound in terms of the location of its instantiation. Sound waves spread out from their sources, and their identity and location at any time can be indeterminate; there may, therefore, be no very determinate answer to the question of where a sound is instantiated. Although it may not be possible to say exactly where a sound is instantiated, the sounds we hear are instantiated by sound waves that affect us; wherever else they may also be instantiated, when we hear them, sounds are instantiated where we are.

The sounds that we hear are instantiated where we are. But they usually are not only instantiated where we are; they usually are instantiated throughout a region of space that includes where we are. Of course that does not mean that we only ever hear part of a sound that affects us: sounds are patterns or structures that are instantiated by sound waves, and the entire pattern or structure is wholly instantiated wherever it is instantiated, including where we are. A sound can be instantiated in different regions of space at different times; therefore, just as the pattern of ripples on the surface of a pool of water can be said to move, sounds can be said to move.[8]

Is this account of sounds inconsistent with our experience of sounds? Several writers have suggested that it is.[9] They claim that we hear the location of the sources of sounds and that we experience sounds as located *at* their sources—they claim that sounds *seem* to be at their sources. If that's right, then any account of what sounds are must—on pain of being committed to a view according to which auditory experience is generally non-veridical—identify sounds with something located at their sources, as events involving their sources or the vibrations of the sources. Whilst a

---

[8] Although sounds can be said to move, we do not experience them as moving (and, indeed, what would be the *point* in hearing sounds move?).

[9] For example, Pasnau (1999), Casati and Dokic (Chapter 5; see also 1994), and O'Callaghan (Chapter 2; see also 2007).

perceptual system may always be susceptible to error, an account that has the consequence that the auditory system is always erroneous is implausible. Since the view that I have outlined entails that sounds are not, or not only, located at their sources, it is, according to this objection, committed to viewing auditory experience as generally non-veridical; as such, it is an implausible view.

A satisfactory response to this phenomenological objection must address two worries. First, can the account that I have given explain how we are able to perceive the location of the sources of sounds? Second, is my account inconsistent with the phenomenology of our auditory experience and committed to viewing auditory experience as generally non-veridical? Before directly addressing those two worries, I want to say something about the role of space in the auditory processes that I have described. Understanding that role will help to answer these questions.

2. Space in Auditory Perception

What role does space play in auditory grouping? Grouping is the process by which the frequency components detected by the ears are grouped so as to correspond to their sources. The fact that a group of components all come from the same location is good evidence that they have been produced by the same source, and the fact that two components come from different locations is good evidence that they were produced by distinct sources. So, we might expect spatial location to play a role in both simultaneous and sequential grouping.

Spatial properties have an influence on grouping, but even when frequency components can be spatially distinguished they are not necessarily grouped according to their spatial properties. Frequency components from the same location are not grouped if other non-spatial cues—such as shared onset times and harmonic relationships—conflict with spatial cues, and frequency components from different locations may be grouped if non-spatial cues indicate that they have the same source. Our experience of music played over stereo loudspeakers is a good example of this. Frequency components produced by spatially distinct sources—the loudspeakers—may be grouped to produce an experience of a single sound, perhaps with an apparent source that is located at a place in between or behind the speakers. This happens because the spatial cues to grouping are weaker than the non-spatial cues: the pattern of frequency components produced by the two speakers is more likely to have been produced by a single sound source whose spatial cues have been disrupted than to have been produced by two sources.[10] Diana Deutsch's 'scale illusion' is an example in which a sequence of sounds is heard as grouped into a sequence despite the fact that individual sounds come from different locations.[11]

Both these examples demonstrate that spatial properties provide only a weak cue to frequency component grouping, a cue which can be overridden by non-spatial cues. This makes ecological sense, given that the auditory system functions to group frequency components that correspond to their sources. The transmission of sound waves—with frequency components being detected only after they have been reflected off and refracted around surfaces—disrupts spatial cues and makes them unreliable. Their unreliability means that other non-spatial cues are a better guide to correct grouping, and that correct grouping according to spatial cues will only reliably correspond to sound sources if those spatial cues are supported by other non-spatial cues.

Although spatial properties can influence auditory grouping, they are not necessary for grouping. Frequency components can be appropriately grouped even when they cannot be spatially distinguished. This is evident from the fact that we can hear sounds produced by different sources as distinct when their frequency components all come from the same location—components are grouped as distinct despite being spatially indistinguishable. This happens, for example, when sounds are played over a single loudspeaker, such as a radio, or when different sound sources are all heard from behind a barrier or wall. Frequency components from sources which are behind one another relative to the listener will tend to have the same spatial cues and so be spatially indistinguishable, yet may be heard as distinct.

The fact that spatial cues are not necessary for grouping should lead us to question in what sense grouping is ever genuinely spatial at all. Frequency components are not themselves intrinsically spatial, nor is the way in which they are detected.[12]

Spatial information about the location of the source of a frequency component must be recovered from time, intensity, and phase differences between

---

[10] By 'more likely' I do not mean true in the closest worlds (it would not be more likely in that sense for someone who heard only artificially produced sounds), but to have been produced that way in the recent evolutionary history of the auditory system.

[11] For a description and discussion of the illusion, see Deutsch (1974, 2004) and for discussion see Bregman (1990: 76).

[12] Neurons of the basilar membrane, which detect properties of the sound wave, are organized in a tonotopic way, unlike the neurons of the retina, which are spatially organized (with the consequence that spatial information is embodied in the spatial pattern of the stimulation of the retina as well as in the binocular differences in the stimulation of the two retinas).

each ear's detection of that frequency component. These relations between the frequency components detected by each ear carry information about spatial location, but they do not have any intrinsic spatial significance. There is no intrinsic connection between the phase and temporal relationships of frequency components and spatial properties. Frequency components only gain spatial significance by being interpreted or encoded in a spatial way.

But such a spatial interpretation or encoding is not necessary for the properties that encode spatial location to play a role in frequency component grouping. Common inter-aural time and phase differences can be used as indications of correct grouping independently of frequency components or groups of frequency components having been encoded within a spatial framework. That means that components do not need to be assigned spatial properties or coordinates and then grouped on the basis of shared or different spatial properties or coordinates; they can simply be grouped on the basis of temporal and phase relationships. If components are grouped on the basis of temporal and phase relationships, then they will be grouped according to whether they come from the same or different locations, but the process of grouping will not be a spatial process—it will not operate on or use spatial properties to determine grouping. We can therefore explain the role that the spatial location of the source of a sound has in grouping frequency components—we can explain how frequency component groupings track sameness and difference in spatial location—without supposing that the processes or grouping are themselves spatial, that they operate on or use spatial properties.

The fact that spatial cues are not necessary for grouping suggests that frequency component groupings are not spatially individuated: distinct groups of frequency components are not distinct in virtue of having different spatial properties. Given that the auditory system's grouping of frequency components determines what sounds we experience, the fact that groupings are not spatially individuated suggests that sounds are not either. It would follow that it is possible to experience two sounds as distinct without experiencing them to be spatially distinct. If my argument is right that grouping is not spatial even when it exploits properties that track spatial differences, then frequency component groupings do not have spatial properties. Again, given that the auditory system's grouping of frequency components determines what sounds we experience, the fact that groupings do not have spatial properties suggests that sounds do not either. It suggests, in other words, that sounds do not have any *intrinsic* spatial significance and do not have any spatial structure.[13]

---

[13] This contrasts with vision. Visual features have spatial properties in virtue of the way they are detected: the retina detects a spatially distributed pattern of light and preserves that spatial mapping

Reflecting on our experience of sounds supports these suggestions about sounds. We often hear two sounds as distinct without their having or seeming to have different spatial properties; we do so when we listen to music on a radio that has a single speaker, or hear sounds coming through the window of the room we are in. In such cases we hear simultaneous sounds as distinct, and can focus our attention on one to the exclusion of the other, without being able to distinguish them spatially. If our attention is focused on a place, it is always the same place; therefore, when we focus our attention on one sound rather than another, we do not focus our attention on one place rather than another. We do not hear sounds as having spatial parts or as having spatial structures. Although we can make sense of a sound having *temporal* parts—in virtue of the fact that we hear sounds as temporally extended—we cannot make sense of *spatial* parts of sounds. Hearing simultaneous sounds as having distinct spatial properties is sufficient to hear them as distinct sounds, so we cannot simultaneously hear distinct parts of a *single* sound as standing in spatial relation to one another. If we do not hear sounds as having spatial parts, then it cannot be that what makes simultaneous parts of a sound seem to be parts of single sound is that they are spatially related, a conclusion that is supported by the fact that it is possible to hear sounds without hearing them as having *any* spatial properties. The same is surely true of the temporal parts of sounds and sequences of sounds. When we hear a melody, we hear a sequence of sounds *as* a sequence. As the example of listening to music on the radio shows, hearing the elements of this sequence as such is not a matter of hearing them as sharing spatial properties.

The claim that sounds are not individuated spatially has empirical support. A soft sound tends to disappear—is masked—if a louder sound of a similar frequency is heard simultaneously; however, if the louder sound is spatially separated from the softer sound, the masking effect is reduced and it is possible to hear the softer sound. Thiran and Clarke (2003) describe a subject, NM, with a spatial hearing deficit (see also Darwin and Hukin 1999). NM can hear sounds and recognize their sources, but cannot localize them or perceive them

through subsequent processing. As a consequence, visual features have an intrinsic spatial significance: they stand in spatial relation to each other without needing to be interpreted spatially (though they may be mapped into other frames of reference for other purposes). There are visual processes—analogous to auditory grouping processes—that bind together different visually detected features as features of a single object. The best explanation of how features belonging to a single object are bound together is that binding processes exploit spatial properties, binding together as features of a single object those features which share spatial properties. Spatial properties are necessary for visual binding, and hence necessary for object identity. The spatial properties that individuate visual objects are properties of the visual field, and are independent of any mapping of visual objects into an egocentric frame of reference.

as moving—sounds presented at different azimuthal positions seem to her all to be at the same position. Despite this spatial deficit, NM experiences the release from masking effect. She cannot hear a soft sound when the masking sound is in the same position as the soft sound, but can hear it when they are spatially separated. When asked to report the location of the masking sound, she said that she always heard it in the same place, superimposed on the softer sound. For this subject, spatial properties play a role in grouping sounds, and so in determining what sounds she experiences, but in a way that does not enter into the content of her experience. She hears sounds as distinct without experiencing them as spatially distinct, and she does not experience sounds as having spatial properties.[14]

Although the processes that determine what sounds we hear are not intrinsically spatial, our auditory experience is, or can be, spatial: we can hear where the sources of sounds are. For us to be able to hear the locations of the sources of sounds, the auditory system must extract spatial information about the locations of the sources from the groups of frequency components that it detects, and from the differences between the groups detected by each ear. If what I have already argued is correct, the auditory system does not interpret individual frequency components as having a spatial significance; it first groups frequency components from a single source, and then extracts from the group spatial information about the source. The process that extracts spatial information occurs as part of the third stage of auditory processing—the stage that functions to extract information about sound sources. This process is distinct from the second stage of grouping processes that I have described. Spatial location is one of a range of properties of the source of the sound, information about which the auditory system extracts from frequency components that have been grouped to correspond to their source.[15]

What kind of spatial information does the auditory system extract? It can only extract what spatial information is available. Spatial information about sound sources is (for the most part) embodied in the differences in phase, time, and intensity of groups of frequency components detected by each ear.[16] This information concerns the location of the source of the sound relative to the

---

[14] Although she describes or indicates the sounds as originating at the same *place*, it is plausible that they do not seem to her to be located or come from anywhere and that she makes her indication only because of the requirements of experiment (in conversation, Clarke has agreed that this is a plausible interpretation of NM).

[15] This suggests an interpretation of the 'what and where' distinction in auditory perception which is at odds with the dominant view in psychology; for an elaboration and defense see my (unpublished b).

[16] Reflections play a role, too, but not in a way that undermines this line of argument (see Blauert 1997: sect. 2).

perceiver's head. The initial interpretation of the location of a sound source will therefore be in a head- or body-centered frame of reference—one that represents the location of the source of a sound relative to the perceiver's head or body. As a consequence of the fact that we perceive the locations of sources relative to where we are, we can, immediately and without calculation, turn our heads or point in the direction of the source of a sound we hear. We also perceive the locations of sound sources relative to one another; we can hear, for example, that one source is to the left of another. Our perceptual system must therefore map the head- or body-centered locations of sound sources into other non-egocentric frames of reference, perhaps frames of reference shared with other senses (Spence and Driver 2004: chs. 10, 11). However, given the way that spatial information is embodied in sound waves, it is possible to map the location of a sound source in a non-egocentric frame of reference only if it has first been located egocentrically. It would seem to be necessary, therefore, that if we experience the source of a sound as having a location, we experience it as having a location relative to where we are.

3. The Spatial Phenomenology of Auditory Experience

The objection to the view of sounds that I outlined at the end of Section 1 is that sounds seem to be located where their sources are. This is a phenomenological claim, a claim which is supported by appeal to how things experientially seem. To answer this objection I will first describe the way in which sound *sources* seem to be located and then ask in what sense, if any, sounds seem to be located *at* their sources.

In his well-known discussion of sight and touch, Martin (1992) draws a contrast between the different ways we experience space and objects as located in space. He points out that in vision, we are aware of a region of space within which we can experience objects—that we experience space itself *and* objects in space—whereas in touch, we experience the location of parts of objects that we are aware of as extended in a space that extends beyond the limits of our experience, but which is not itself experienced. He illustrates the contrast with an example:

Consider the case of looking at a ring-shaped object, a Polo mint, for instance, head on. One is aware of the various white parts of the mint arranged in a circle, and aware of how they are related to each other. One is also aware of the hole in the middle of the mint, and that that hole is there in the middle. If one was not aware of the hole one would not see the mint as ring shaped rather than a circle. Nothing need be perceived to be within the hole. One is aware of the hole as a place where something potentially could be seen, not as where something is actually seen to be... So we can think of normal visual experience as experience not only of objects which are located in some space, but as of a space within which they are located. The space is part of the experience in as much as one is aware of the region as a potential location for objects of vision.   (Martin 1992: 199)

This description of an experience as *of* space does not seem an appropriate description of our tactile experience:

... when one grasps the rim [of a glass] one comes into contact with it at only five points, where one's fingertips touch it. Nevertheless one comes to be aware that the glass as a whole is circular. In being tactually aware in this way, is one aware of the parts of the rim in between the points of contact in the same way as one is aware of those points, and is one aware of the region of space lying inside the rim? The answer would appear to be not: one comes to be aware of the glass by being aware of the parts one touches. In this it contrasts with the Polo mint, since one is aware both of the ring-surface and the hole in the same way.   (Martin 1992: 200)

The description of visual experience as an experience *of* space does not seem appropriate as a description of our auditory experience either. Suppose that you hear two sound sources as located one in front and to the left of you, and one in front and to the right of you. You are aware of the locations of the sources and aware that they are a certain distance apart, separated by a region of space. Your awareness of these places is similar to the tactile awareness you might have of the rim of a glass. Just as in the tactile case, where we are aware of the rim at the points we touch it and there is a contrast between our awareness of the locations of the points of contact and our awareness of the space between those points, in the auditory case you are aware of places in virtue of hearing something to be located there, and there is a contrast between the way in which you are aware of the places where you hear something to be and your awareness of the region of space that separates those places. In both touch and hearing, the space that separates the experienced locations is not itself an object of the experience. In neither case are we aware of the region of space between the places we experience something to be, in the way that we are visually aware of the empty space that is the hole in the Polo mint. In this respect, then, the phenomenology of our auditory experience of space is more similar to the phenomenology of our tactile than to our visual experience of space.

But this poses a puzzle. In order to mark the phenomenological differences between our visual, tactile, and auditory experiences of things in space, we seem compelled to say that although in vision we are aware of space, this is not true of touch or hearing. In touch and hearing, we are aware of things in space, but we are not aware of space itself. However, there is a sense in which our auditory and tactile experiences *do* provide us with an awareness of space. Although we only touch the rim of a glass at five points, we are aware of the rim as circular and so as occupying the space in between the points we touch; although we hear the location of two objects relative to one another, we are aware of them as separated in space and so aware of the space in between where we hear them to be. On the one hand, in order to mark the phenomenological differences, we need to deny that we are auditorily and tactilely aware of space. On the other hand, we can only explain how we are auditorily and tactilely aware of things in space by supposing that we have both an auditory and a tactile awareness of space. So we both do and do not have a tactile and auditory awareness of space.

The problem is to explain the phenomenological contrast, in tactile and auditory experience, between our awareness of the places we experience something to be and our awareness of the places where we do not experience anything to be (those places, for example, in between the places we experience things to be), and so to give some account of how we are aware of places we do not experience anything to be.

Prima facie, this does not seem possible within the framework of a representational theory of perception. A representational theory of perception explains what it is to experience something in terms of the representational content of an experience. We have an experience of something in virtue of our experience representing it as being some way. To say that we are experientially aware of something is then just to say that we are aware of it in virtue of having an experience of it. But that means we cannot draw a distinction between experiencing something and being experientially aware of it. We can either say that we are aware of space in virtue of our experience representing that space, or that we have an experience of space in virtue of our experience representing it. Both amount to the same thing. How then do we draw the contrast we need to explain the differences between visual, tactile, and auditory experience of space?

The suggestion that I want to explore is that we can explain the contrast in terms of the determinacy with which spatial properties are represented by different experiences. What does it mean to say that something is represented more or less determinately? I have in mind the intuitive idea that an experience can tell us more or less about some region of our environment by representing that region and the objects and properties within it in greater or less detail, in a more or less determinate way.

We are familiar with the idea that pictures can represent more or less determinately. A color photograph of an apple tells us more about the apple than a black and white photograph does; the black and white photograph does not tell us what color the apple is—whether it is red, for example. A black and white photograph neither represents that the apple is red nor that the apple is not red. We should not think of a black and white photograph as *mis*representing the color of the apple—as representing it to have a color it does not have—it is simply silent on the question of the apple's color and so does not represent the apple as having a determinate color. Similarly, an outline drawing of a bird is in many respects indeterminate. It does not represent the color of the bird, nor whether the bird has feathers, nor whether the surface of its back is different to that of its legs. Again, it is not that the drawing is misleading; it is just not very informative. It does not tell us much about the bird—whether it has or lacks certain properties; it is, in certain respects, indeterminate.

This idea of a representation representing in more or less determinate ways can be applied to perceptual experience. A perceptual experience represents the perceiver's environment as being some way in virtue of having representational content. We can specify the representational content of a perceptual experience by specifying those ways which the perceiver's environment can be that are consistent with the representational content of the experience being correct. Thought of this way, the representational content of an experience specifies a set of ways the perceiver's environment can be, and the experience is veridical only if the perceiver's environment is one of those ways. The same approach can be used to specify the way in which an experience represents an object: we specify which ways the object can be that are consistent with the representational content of the experience being correct. A perceptual experience whose content is relatively determinate will specify a narrower set of ways that things could be; by telling us more about how things are, it narrows down the possible ways things could be. An experience whose content is relatively indeterminate will specify a wider set of ways that things could be; by telling us less about how things are, it allows a greater number of possible ways things could be. We can use this notion of the indeterminacy of experience to explain the differences between visual and tactile experiences of space.

For example, a visual experience that represents the rim of a glass as circular is more determinate than a tactile experience that represents the rim of a glass as circular; the visual experience narrows down the ways a certain region of space could be far more than the tactile experience. It does so because it tells us more about the rim, the glass, and the space around the glass. It represents the rim of the glass as circular and for a region of space—the region that is visible from the subject's point of view—it represents, for every location within that region, either something as at that location or that location as empty. The experience is veridical only if the visible region of space is occupied at those locations where something is represented to be, and empty at those locations where nothing is represented to be. It is because it represents every location within a region of space as determinately occupied or empty that visual experience is an experience of space as well as objects in space. A tactile experience of the rim of the glass represents the rim of the glass as circular, and so represents the rim as occupying a region of space that extends between the points of contact with the glass, but is silent about the larger space within which the rim of the glass is experienced to be. It does not represent any location within that larger space as either occupied or as empty. That means that there are many ways that larger space could be that are consistent with the experience being veridical. There could, for example, be a circular piece of opaque card stuck just inside the rim of the glass. The tactile experience does not rule out such a possibility, whereas the visual experience does. By representing the rim of the glass as occupying a larger region of space, the tactile experience represents spatial locations that are not locations of parts of the rim; however, unlike visual experience, it does not tell us anything about those locations—it does not represent them as occupied or as not occupied—and so does not provide us with an experience *of* the space. As a consequence of these differences, there are far fewer ways that the region of space could be that are consistent with the visual experience being veridical than are consistent with the tactile experience being veridical. Similarly, the visual experience of the rim of the glass represents each part of the rim in an equally determinate way. In contrast, the tactile experience of the rim of the glass represents the parts of the rim at the points of contact more determinately than the parts in between the points of contact. We are aware of the texture of the rim at the points of contact, for example, but not at the points in between. Therefore the experience tells us less about the parts of the rim in between the points of contact than it does about those points; this difference explains the contrast between our awareness of the rim at the points of contact and our awareness of the points in between. Although both visual and tactile experiences tell us that the rim is circular, because it tells us more about the rim of the glass and the surrounding space, there are fewer ways the rim could be that are consistent with the visual experience than with the tactile experience being correct.[17]

---

[17] Amodal completion provides the closest visual analogue of our tactile experience of shape. The fragmentary outline of an occluded circle may still be seen as circular—the fragments are experienced

This explains the different ways in which we are aware of space in touch. A tactile experience tells us about the places we experience something to be—that they are occupied—and it tells us that there are places in between the places we experience something to be; but it tells us nothing about those in-between places. It does not tell us, for example, whether or not they are occupied. There is, therefore, a contrast between our awareness of the places we experience something to be and our awareness of the places in between.

We can explain the phenomenological differences between vision and touch, in particular the differences in our experience of space, by appealing to the determinacy of the content of visual and tactile experiences. We can explain the phenomenology of our auditory experience of space in a similar way.

Our auditory experiences represent space in a way that is often far less determinate than either visual or tactile experience. This is most obviously true of an experience of a sound source that does not seem to be located anywhere. Such an experience is indeterminate with respect to the location of the source. It does not tell us anything about where the source is located and its veridicality is independent of the actual location of the source. When we do experience the source of a sound as located, our experience may tell us more or less about where the source is. We may experience a source as being outside of the room, as being somewhere over on the right, or as being over on the right and just in front of us. In each case the experience veridically represents the location of the source just in case it is somewhere outside the room, or somewhere to our right, or somewhere to our right and in front. The experiences differ in the determinacy with which they represent the location of the source; greater determinacy corresponds to smaller regions of space within which the source must be located if the experience is to be veridical. Our auditory spatial acuity is relatively poor, so even in the best circumstances the most determinate auditory experience is quite indeterminate about spatial location.

Auditory experience does not tell us anything about regions of space other than those where we experience sound sources to be. If we experience the source of a sound to be over on our right and another source to be over on our left, then we are aware of their being separated by a region of space, and our experience represents the spatial relation between them. But our

---

as parts of a circle—but there is a contrast between our visual awareness of the visible fragments and our awareness of the parts of the circle that are occluded. When we see the letter 'B' obscured behind a squiggle, we experience only parts of the letter, but are aware of the surface as forming the letter 'B'. Our experience represents the letter as shaped like a 'B' and so represents the surface as extending underneath the squiggle. But we do not *experience* the surface underneath the squiggle; we just experience those parts of the letter that are not obscured.

experience does not tell us anything about that space; in particular, it does not tell us whether there is anything at the places in between the places we hear the sources to be. It does not represent those locations as either occupied or as empty.

Whilst it is true that we experience the location of the sources of sounds, and that sound sources seem phenomenologically to be located, simply saying that sound sources seem to be located obscures the differences between the way auditory and visual experience represents the spatial location of objects. Sound sources seem to be located in a way that is different from the way objects that we see seem to be located. In particular, sound sources seem to be located only in virtue of seeming to be somewhere relative to where we are and to other sound sources. Our auditory experience represents sound sources as located in space in virtue of representing them as standing in spatial relation to one another and to us.

Unlike visual experience, auditory experience does not represent empty places—it does not represent places as unoccupied. That means that auditory experience only represents spatial relations between the locations where we experience the sources of sounds to be, and between those places and us. Visual experience represents both the locations of objects and of places as unoccupied, and it represents spatial relations both between the locations of objects and between those objects and places it represents as unoccupied. Auditory experience does not represent places as unoccupied and does not, therefore, represent the spatial relations between the locations of sound sources and of unoccupied places. Visual experience represents objects and parts of objects, and the spatial relations between objects and their parts. Auditory experience does not represent sound sources as having parts and so does not represent parts of objects as spatially related to one another. Unlike our visual experience, our auditory experience of space is exhausted by our awareness of spatial relations between sound sources and us, and between sound sources and other sound sources. Our visual awareness of space is not exhausted by our awareness of objects as spatially located relative to each other and to us. Visual experience represents the spatial properties of objects, and it represents objects as having parts that are spatially related to one another. Visual experience represents places, whether occupied or not, and so represents spatial relations between places and objects. We are visually aware of the spatial relations of objects to space itself—to places where we experience nothing to be.

There are, then, significant differences between our visual experience of the location of objects and our auditory experience of the location of objects, and so between the sense in which objects we see *seem* to be located and the sense in which the objects we hear *seem* to be located, and it does not follow from the fact that sound sources seem to be located that they seem to be located in the same way that visual objects seem to be located.

We experience the sources of sounds as located. Do we experience sounds as located? Sounds do not seem to be located anywhere other than at their sources; therefore, if sounds seem to be located, they must seem to be located where their sources are. To answer the question, then, of whether sounds seem to be located, we need to answer the question of whether we experience sounds as located where we experience their sources to be. When we hear the alarm clock ringing, we can hear where the clock is—that it is on our left-hand side. Do we also hear where the sound of the alarm clock is? In other words, does the sound of the alarm seem to be where the alarm clock seems to be?

One model for how sounds might seem to be located is a visual one. We can see a pattern or mark as located where an object is located by seeing the pattern as a pattern on the surface of the object. Suppose, for example, that we see a cube with a cross marked on one of its faces. We can see where the cube is—we experience the cube as having a location—and we experience the cross as located where we experience the cube as located. The cross seems to be where the cube is because it seems to be on the surface of the cube, and the cross seems to be on the surface of the cube—at least in part—because it has spatial parts that we experience as spatially related to parts of the surface of the cube. This explanation of the cross seeming to be where the cube is depends on both the pattern and the cube having spatial parts, and our experiencing parts of the pattern as spatially related to parts of the cube.

In this example, the pattern and the object can both be identified independently of one another, and each has a spatial location. In seeming to see the pattern as on the surface of the cube, we seem to see it as having the *same* location as that of the surface. Its having the same location as the cube is a contingent matter. We could see the pattern as being located elsewhere, and in seeming to see the pattern as on the surface of the cube, our experience may mislead us. The pattern could be located elsewhere and merely appear, as the result perhaps of an arrangement of mirrors, to be located on the surface of the cube.

This model, of two independently identifiable objects which can differ in location but which are experienced as having the same location—as sharing spatial properties—is the only one according to which it could follow from the *spatial* phenomenology of experience that one object has the same location as another object. Each object seems to have a location and both seem to have the same location. It is not a model that applies to our auditory experience of sounds and their sources. We do not experience sounds as having a spatial location independently of their sources having a spatial location, nor do we experience sounds as having spatial parts, so we do not experience sounds as having the same spatial location as their sources, nor as having parts that stand in spatial relation to parts of their sources. If sounds seem to be located at their sources, it is not *because* they seem to have the same location as their sources.

Can sounds seem to be where their sources are other than in virtue of seeming to share spatial properties with their sources? Perhaps, by analogy with color, they can. The surfaces of objects appear colored. When we see the surface of an object as colored, the color seems to be where the surface is. The color does not seem to be where the surface is in virtue of seeming to share spatial properties with the surface; it does so because it seems to be a property or quality *of* the surface. Because of the way they seem, any account of what colors are that claims that colors are not properties of surfaces must view the phenomenology of our experience of colors as misleading, and so is committed to an error theory of color experience of the kind proposed by Boghossian and Velleman (1989).

The phenomenological objection to my account of sounds may best be understood in a similar way: sounds seem to be properties or qualities of the objects that are their sources and so seem to be located where those objects are located—not in virtue of sharing spatial properties with those objects, but in virtue of seeming to be properties or qualities *of* them. It would follow that any view of sounds that claims that sounds are not properties of objects and not located where those objects are located must view the phenomenology of our experiences of sounds as misleading, and so is—unacceptably—committed to an error theory of auditory experience. If that's right, then it is not that my account of auditory experience gets the *spatial* content of that experience wrong; it fails to accommodate the fact that sounds seem to be qualities of objects.

Is an objection along these lines right? Do the sounds we experience seem to be properties of their sources in the same way that colors seem to be properties of the surfaces of objects? To answer that question we need first to determine what it is about our experience of colors that makes it correct to describe them as seeming or appearing to be properties of the surfaces of objects. Although it is widely agreed that colors *do* seem that way, what it is about how they seem that makes that description appropriate is not often discussed. There are, nonetheless, features of the appearance of colors that make that description appropriate.

The way the color of a surface appears is determined by both the color of the surface and the way that the surface is illuminated. The way a surface appears—in particular the way the color of a surface appears—changes according to how the surface is illuminated. For example, a surface of uniform color that has a shadow cast on it has a different appearance to a surface without a shadow cast on it. Such differences in the way the color of a surface appears do not normally appear to be differences in the color of the surface; they appear to be what they are, differences in the way the surface is illuminated.[18] Normally, we can distinguish in our experience between differences in how the color of a surface appears which are due to variations in the color of the surface, and differences in how the color of a surface appears which are due to variations in how the surface is illuminated.

Our ability to make this distinction in our visual experience depends on the fact that the color of a surface can appear to be constant through changes in illumination. To see a shadow on a surface as a shadow is to see it as a discontinuity in the illumination of a uniformly colored surface rather than as discontinuity in the color of a uniformly illuminated surface. (Conversely, to see a discontinuity as a discontinuity in the color of the surface is to see a change in the way the surface appears as independent of its illumination.) We cannot explain the appearance of the color of the surface other than in terms of the way two things appear to interact: the color of the surface appears to interact with the light illuminating it.

It is because we can distinguish between those changes in the way the color of the surface appears that are due to variations in the apparent color of the surface, from those that are due to apparent variations in the way the color is illuminated, that it is correct to describe colors as seeming to be surface properties of objects—as properties of surfaces that stay constant through changes in illumination and which partly determine how the surface appears. If we could not make this distinction, then colors would not seem to be properties of the surfaces of objects. Sometimes (when, for example, a surface is viewed through a reduction screen that prevents the perception of its illumination), it is not possible to tell whether a change in the appearance of a color is a variation in the surface or a variation in the illumination of the surface. In such cases, the color does not seem to be the color of the surface,

---

[18] That they appear this way is not the result of our *judging* that a variation is a variation in illumination rather than a variation in the surface; it is a matter of how they experientially appear. That variations in illumination appear differently to variations in surface color is a consequence of the operation of color and lightness constancy mechanisms in visual processing. See Gilchrist (2006) for a state-of-the-art discussion of the mechanisms of lightness constancy.

but has the appearance of what Katz called 'film' color—color that appears transparent and is looked through rather than at.[19]

Are there similar features of the appearance of sounds that would make the description of sounds as apparent properties of objects appropriate? The appearance of the sounds we hear is determined both by the character of the vibration of the object that produced the sound and by the way that vibration is altered during its transmission from its source to our ears. To justify the description of sounds as seeming to be properties of the objects that produce them, there should be a similar pattern within our experience of sounds to that within our experience of colors. The way sounds appear should be explained as the result of the apparent interaction of the sound-of-the-object and the transmission of the sound. That is, the way sounds appear should appear to be the result of the interaction of the-sound-of-the-object and the way the-sound-of-the-object interacts with its environment. Then sounds would appear to be properties of their sources similarly to the way that colors seem to be properties of surfaces.

The appearance of a sound does not *appear* to be the joint upshot of the apparent sound-of-an-object together with the apparent alteration of the sound during transmission. That is, we cannot distinguish *in experience* between aspects of how sounds appear which are due to the sound-of-the-object, and aspects of how sounds appear which are due to alterations of the sound during transmission. The appearance of a sound is simply determined by how the sound we hear appears to be. (Contrast this with color. The appearance of a color—for non-film colors—is not simply determined by how the color appears to be, but by how the color appears to be together with how the color appears to be illuminated.)

We can, of course, distinguish in auditory experience between how a sound appears to be and how the source of that sound appears to be. How a sound appears can change without the source of that sound appearing to change. A sound may appear to be quieter, for example, without the event that produced it appearing to change. When we close the window to shut out the sounds of the traffic, the sounds we hear get quieter, but what is making the sounds does not appear to change. When you pull your hat down over your ears, the sounds you hear become muffled, but the things producing those sounds appear unaltered. What appear not to change are the objects that produce the sounds, not the sounds that they produce. The sounds they produce *do*

---

[19] Katz (1935: 10) wrote that 'a complete impression of illumination is had only where objects are perceived, and...whenever objects are perceived an impression of illumination is always produced'.

appear to change—they appear to be quieter or muffled. It is not that the sound-in-the-object appears to be unchanged through changes in how the sound appears; the object appears to be unchanged through changes in how the sound appears to be. The sound-in-the-object does not appear to be any way at all.

There are no grounds, therefore, for our saying that the sound-in-the-object appears responsible for how the sound appears in the way that the surface color of an object appears responsible for how the color of the surface appears; and no grounds, therefore, for saying that sounds seem to be properties of objects rather than things produced by objects. In fact, the opposite is true. That sounds can appear to change, and can be changed without their objects appearing to change, is what leads us to describe sounds as seeming to *come from* their sources, and makes such a description appropriate.[20]

Sounds do not seem to be at their sources in virtue of sharing spatial properties, nor in virtue of seeming to be qualities of their sources. Therefore, the phenomenology of auditory experience is not inconsistent with an account of sounds, such as mine, according to which sounds are not located at their sources. In fact, the opposite is true: the account I have given is not only consistent with that phenomenology, it provides the best explanation of it.

References

Blauert, J. (1997). *Spatial Hearing: The Psychophysics of Human Sound Localization*, rev. edn. Cambridge, Mass.: MIT Press.

Boghossian, P. A. and Velleman, J. D. (1989). 'Color as a Secondary Quality'. *Mind*, 81–103.

Bregman, A. S. (1990). *Auditory Scene Analysis: The Perceptual Organization of Sound*. Cambridge, Mass.: MIT Press.

Cabe, P. and Pittenger, J. B. (2000). 'Human Sensitivity to Acoustic Information from Vessel Filling'. *Journal of Experimental Psychology: Human Perception and Performance*, 26: 313–24.

Carello, C., Anderson, K. L., and Peck, A. (1998). 'Perception of Object Length by Sound'. *Psychological Science*, 9: 211–14.

---

[20] Note that although sounds appear to change independently of objects, the same is *not* true of colors. The appearance of the color of a surface can change independently of the surface of the object appearing to change, but such a change is not a change in the apparent color of the surface; it is a change in the apparent illumination of the surface. If the apparent illumination is held fixed, then the appearance of the color of the surface can only change if the color of the surface—and so the object—appears to change.

——Wagman, J. B., and Turvey, M. T. (2005). 'Acoustic Specification of Object Properties', in J. D. Anderson and B. Fisher (eds.), *Moving Image Theory: Ecological Considerations*. Carbondale, Ill.: Southern Illinois University Press.

Casati, R. and Dokic, J. (1994). *La Philosophie du Son*. Nîmes: Chambon.

Darwin, C. J. and Hukin, R. W. (1999). 'Auditory Objects of Attention: The Role of Interaural Time Differences'. *Journal of Experimental Psychology: Human Perception and Performance*, 25: 617–29.

Deutsch, D. (1974). 'An Auditory Illusion'. *Nature*, 251: 307–9.

——(2004). 'The Octave Illusion Revisited Again'. *Journal of Experimental Psychology: Human Perception and Performance*, 30: 355–64.

Fletcher, N. H. and Rossing, T. D. (1998). *The Physics of Musical Instruments*, 2nd edn. New York: Springer.

Fowler, C. A. (1991). 'Auditory Perception is Not Special: We See the World, We Feel the World, We Hear the World'. *Journal of the Acoustical Society of America*, 89: 2910–15.

Freed, D. J. (1990). 'Auditory Correlates of Perceived Mallet Hardness for a Set of Recorded Percussive Events'. *Journal of the Acoustical Society of America*, 87: 311–22.

Gaver, W. W. (1993a). 'How Do We Hear in the World? Explorations in Ecological Acoustics'. *Ecological Psychology*, 5: 285–313.

——(1993b). 'What in the World Do We Hear? An Ecological Approach to Auditory Event Perception'. *Ecological Psychology*, 5: 1–29.

Gilchrist, A. (2006). *Seeing Black and White*. New York: Oxford University Press.

Kaplan, D. (1990). 'Words'. *Proceedings of the Aristotelian Society Supplementary Volume*, 64: 93–119.

Katz, D. (1935). *The World of Color*. London: Kegan Paul, Trench, Trubner, and Co.

Kunkler-Peck, A. and Turvey, M. T. (2000). 'Hearing Shape'. *Journal of Experimental Psychology: Human Perception and Performance*, 1: 279–94.

Lakatos, S., McAdams, S., and Caussé, R. (1997). 'The Representation of Auditory Source Characteristics: Simple Geometric Form'. *Perception and Psychophysics*, 59: 1180–90.

Li, X., Logan, R. J., and Pastore, R. E. (1991). 'Perception of Acoustic Source Characteristics: Walking Sounds'. *Journal of the Acoustical Society of America*, 90: 3036–49.

Lutfi, R. A. and Oh, E. (1997). 'Auditory Discrimination of Material Changes in a Struck-Clamped Bar'. *Journal of the Acoustical Society of America*, 102: 3647–56.

McAdams, S. (1993). 'Recognition of Sound Sources and Events', in S. McAdams and E. Bigand (eds.), *Thinking in Sound*. Oxford: Oxford University Press.

Martin, M. G. F. (1992). 'Sight and Touch', in T. Crane (ed.), *The Contents of Experience*. Cambridge: Cambridge University Press.

Neuhoff, J. (2004). 'Auditory Motion and Localisation', in J. Neuhoff (ed.), *Ecological Acoustics*. London: Academic Press.

Nudds, M. (unpublished a). 'Auditory Perception'. <http://homepages.ed.ac.uk/mnudds/auditory.html>.

Nudds, M. (unpublished b). 'What and Where in Auditory Perception'. <http://homepages.ed.ac.uk/mnudds/auditory.html>.

O'Callaghan, C. (2007). *Sounds: A Philosophical Theory*. Oxford: Oxford University Press.

Pasnau, R. (1999). 'What is Sound?' *Philosophical Quarterly*, 49: 309–24.

Russell, M. and Turvey, M. T. (1999). 'Auditory Perception of Unimpeded Passage'. *Ecological Psychology*, 11: 175–88.

Schiff, W. and Oldak, R. (1990). 'Accuracy of Judging Time to Arrival: Effects of Modality, Trajectory, and Gender'. *Journal of Experimental Psychology: Human Perception and Performance*, 16: 303–16.

Spence, C. and Driver, J. (2004). *Crossmodal Space and Crossmodal Attention*. Oxford: Oxford University Press.

Thiran, A. B. and Clarke, S. (2003). 'Preserved Use of Spatial Cues for Sound Segregation in a Case of Spatial Deafness'. *Neuropyschologia*, 41: 1254–61.

VenDerveer, N. J. (1979). 'Ecological Acoustics: Human Perception of Environmental Sounds'. Ph.D. thesis. Dissertation Abstracts International, 40, 4543B. (University Microfilms No. 80-04-002).

Warren, R. M. (1999). *Auditory Perception: A New Analysis and Synthesis*. Cambridge: Cambridge University Press.

Warren, W. H. and Verbrugge, R. R. (1984). 'Auditory Perception of Breaking and Bouncing Events: A Case Study in Ecological Acoustics'. *Journal of Experimental Psychology: Human Perception and Performance*, 10: 704–12.

—— and Whang, S. (1987). 'Visual Guidance of Walking through Apertures: Body-Scaled Information for Affordances'. *Journal of Experimental Psychology: Human Perception and Performance*, 13: 371–83.

Wildes, R. and Richards, W. (1988). 'Recovering Material Properties from Sound', in W. Richards (ed.), *Natural Computation*. Cambridge, Mass.: MIT Press.

5
Some Varieties of Spatial Hearing[1]

ROBERTO CASATI AND JÉRÔME DOKIC

1. Introduction

A principle of classification of metaphysical theories of sounds can be based on the alleged location each theory assigns to sounds. Sounds can be said to be located either where their material sources are (giving rise to a family of *distal theories*), or where the hearer is (*proximal theories*), or somewhere in between (*medial theories*). In Casati and Dokic (1994, 2005), we argued that a major shortcoming of proximal and medial theories, as opposed to distal theories, is that they do not locate sounds where an untutored description of what is heard suggests they are, namely at their sources. As a consequence, these theories face the obligation of providing an explanation of why auditory perception allows for such a massive error. Then, confident that we stood on phenomenology's side, we put forward our own version of a distal theory, the Located Event Theory, according to which sounds are physical events located (and normally heard as located) where their sources are.

Of course this principle of classification leaves out other theories, in particular what we may dub '*a-spatial theories*'. According to a-spatial theories of *sounds*, sounds do not really have locations in physical space. They are neither individuated in spatial terms nor located anywhere. A-spatial theories of sounds are neither distal nor proximal nor medial, since they invite us to deny that sounds are spatially located entities in the first place. At first sight, our phenomenological argument is still valid against these theories. For doesn't one usually *seem* to hear sounds to be located somewhere in space, namely at their sources? Again, a-spatial theories seem to be saddled with a commitment to an error theory of auditory perception.

---

[1] Research for this chapter was partly funded by the European Commission, Sixth PCRD Network of Excellence 'Enactive Interfaces', IST-2002-002114.

However, consider a-spatial theories of *auditory perception*. The claim here is not that sounds are non-spatial, but that one does not really *hear* them to be located anywhere. The only things that the hearer can locate on the basis of her auditory perception are not the sounds themselves, but (at best) their sources; she hears the trumpet, rather than the sounds it makes, to be located somewhere in the distance. A-spatial theories of auditory perception are in principle independent from a-spatial theories of sounds. For instance, one may argue that sounds themselves are not heard to be located even though they actually are somewhere in physical space.

A-spatial theories of auditory perception threaten our phenomenological argument against proximal, medial, *and* a-spatial theories of sounds. For if one does not hear sounds to be located anywhere, auditory experience is *neutral*—rather than massively in error—as far as the locations of sounds are concerned.

In what follows, we provide some meta-theoretical constraints for the evaluation of a-spatial theories of sounds and auditory perception. We shall point out some forms of spatial content auditory experience can have. Our tentative conclusion is that if auditory experience does not necessarily have a rich *egocentric* spatial content (the kind of content that enables the hearer to locate sources in her egocentric space—for instance, to the left and far away), it must have *some* spatial content for the relevant mode of perception to be recognizably auditory. An auditory experience devoid of any spatial content, if the notion makes sense at all, would be very different from the auditory experiences we actually enjoy. This is enough to dismiss current a-spatial theories of auditory perception. As a consequence, our initial taxonomy of proximal, medial, and distal theories, as well as our phenomenological argument in favor of distal theories, are still topical.

2. The Located Event Theory

Let's start with articulating our own view, the Located Event Theory.[2] According to this view, sounds are monadic events happening to material objects. This means that sounds are *located at their sources*, and are identical with, or at least supervene on, the relevant physical processes in them. This in turn means that:

---

[2] This section elaborates on part of Casati and Dokic (2005).

(i) The relevant physical processes in the sounding object do not move any more than sounds appear to.
(ii) They do not propagate from the object, just as sounds do not appear to.
(iii) Like sounds, and unlike sound waves in the ambient medium, their intensity can remain the same through a period of time, even if one distances oneself from the source and hence hears them as less loud.
(iv) Finally, tuning forks and other sounding objects can be taken as continuing to vibrate irrespective of their being or not being immersed in a medium. We do not create sounds by surrounding vibrating objects with a medium (for instance, air)—we simply *reveal* them.

These four features of sounds construed as located events are in agreement with the phenomenological description of sounds.

Prima-facie objections to the Located Event Theory tend to either misconstrue the phenomenology or beg the question against the idea that sounds are located. For instance, it may be argued that echoes provide a counter-example to the Located Event Theory insofar as the sound as located event is not where phenomenology has it. However, this is a clear case of misrepresentation, comparable to that of seeing an object in a mirror. Missing this fact would lead right into Hobbes's sophism (Hobbes 1651/1839: I, I) that colors are not in things because we can 'sever' them from things by using a mirror, which, generalized, leads to the awkward idea that material objects are not where we see them.

Another alleged problem for the Located Event View is the fact that many features of sounds as they are heard are explained by medial or proximal properties, not by distal properties. This is for instance said to be the case with the Doppler Effect. However, the Located Event View is able to claim that when we experience the Doppler Effect, we just hear the sound in the sounding object, whose features are distorted by the particular experiential situation (relative movement of the sounding object and the hearer).

Furthermore, the Located Event Theory does not make sound perception impossible. Auditory perception of sounds requires a medium which transmits information from the vibrating object to the ears; however, what occurs in the transmitting medium is not constitutive of sounds. One may consider a simple analogy with light. Light is causally responsible for the perception of an object's surface. But this does entail that light itself is seen. The same, according to the Located Event Theory, holds for medial sound waves. Medial sound waves are necessary for perception, but what is perceived is not those waves, but distal sounds.

Finally, one may consider alternative phenomenological descriptions that appear to be more in line with either proximal or medial theories of sound. One may claim that the impression of having a sound in one's ear (purely subjective sound) is enough to favor a proximal description. More interestingly, one may want to put some weight on the alleged meaningful use of expressions such as 'the sound fills the room', and 'sounds fill the room', which appear to speak in favor of an alignment of phenomenology on the medial conception. But we just question the fact that this is an adequate phenomenological description; it may be simply the projection of a true epistemological claim (the fact that the sound is reckoned to be audible from any place in the room) onto a false claim about perceptual content. In this respect sounds are unlike fog, which can literally be seen to fill a room.

3. Assessing the Located Event Theory

We think that the Located Event Theory is superior to its competitors among distal theories. One such competitor is the Relational Event Theory propounded by O'Callaghan (Chapter 2; 2002, 2007), which claims that sounds are events involving both the source and the surrounding medium. They are relational rather than monadic events.

The Relational Event Theory appears to rely upon an argument from vacuums; that is, the fact that sounds are not heard in a vacuum. Of course the Located Event Theory is not affected by a metaphysical reading of this claim, as it does recognize that medial waves are necessary for hearing—if you put a bell in a vacuum jar as the bell goes on vibrating, the sound is still present, although it is not heard. On the other hand, we find the phenomenological reading of this claim questionable in the light of the existence of auditory analogs of tunnel effects, in which unheard items can be still perceptually present. If you could instantaneously empty and refill the vacuum jar in swift alternation, you would hear the sound of the vibrating bell not as going in and out of existence, but simply as not being audible.

We discussed the relative merits of the varieties of distal theories in Casati and Dokic (2005). At this stage, it looks as if the Located Event Theory is a simpler theory which does more justice to the representational power of auditory perception. Let's now see how the Located Event Theory fares with respect to a-spatial theories of sounds and auditory perception.

4. Strawson's Thought-Experiment

A-spatial views of sounds, in more or less strong versions, are widespread in the psychological and philosophical literature.[3] These views often lead to a-spatial theories of auditory perception. For instance, Strawson's famous thought-experiment of a no-space world whose inhabitant's only sensory modality is auditory is motivated by the claim that sounds are intrinsically non-spatial. Here is what he writes:

> Where experience is supposed to be exclusively auditory in character, there would be no place for spatial concepts... Sounds... have no intrinsic spatial characteristics: such expressions as 'to the left of', 'spatially above', 'nearer', 'farther', have no intrinsic auditory significance... A purely auditory concept of space... is an impossibility. The fact that, with the variegated types of sense-experience which we in fact have, we can, as we say, 'on the strength of hearing alone' assign directions and distances to sounds, and things that emit or cause them, counts against this not at all. For this fact is sufficiently explained by the existence of correlations between the variations of which sound is intrinsically capable and other non-auditory features of sense-experience. (Strawson 1959: 65–6)

This passage involves two distinct claims: a *metaphysical* claim according to which sounds are not spatial entities, and an *epistemic* claim according to which there can be auditory experiences devoid of any spatial content. The former claim leads to the latter in the sense that if one wants to avoid an error theory of auditory experience, one must explain away the spatial content of such experience. According to Strawson, the spatial content of *ordinary* auditory experience is fixed by non-auditory features of sensory experience. A *purely* auditory experience, by contrast, would not be one in which sounds would appear to be located in egocentric space.

Given the essential multimodal nature of perception, the notion of a purely auditory experience is suspect. The spatial content of normal auditory perception arguably depends on other senses. It also depends on intentional action, not least because locations in egocentric space are also locations in behavioral space (Evans 1986). However, it is plausible that the dependence among audition, other sensory modalities, and action is constitutive rather than just causal. In this case, if auditory experience is intrinsically spatial, it is not clear that Strawson avoids an error theory after all.

It might be objected that we should interpret Strawson as trying to ground the metaphysical claim on the epistemic claim, rather than the other way

---

[3] Discussed among others by authors like Lotze, Binet, Heymans, Stumpf, Wellek, Révész, Strawson, and Evans. See Casati and Dokic (2005).

round. However, as with all thought-experiments, we should be cautious in moving from an epistemic possibility to a metaphysical possibility.

Let's accept that one can hear distinct sounds even when one does not know where they are in space relative to each other or relative to oneself (this is also true in some pathological cases, as we shall show later). For instance, if one hears certain types of sounds underwater, one has the impression of perceiving sounds without definite spatial localization (aural disparity is insufficient for spatial discrimination, as sound travels at about 1500 m/s in water, more than four times faster than in air, and as our hearing is adaptively tuned to air). Underwater sounds appear to 'just happen', without being heard as being anywhere. The behavioral response witnesses this. One does not spontaneously turn one's head towards the source. All one can do, in order to find the source, is to randomly move around, trying to detect differences in intensity, so as to incrementally approach the source.

The fact that we can imagine a non-spatial auditory experience does not immediately justify the conclusion that there can be a world exclusively populated with sounds, here construed as entities without any spatial structure. Here are more focused formulations of the relevant claims:

1. I imagine hearing sounds independently of an egocentric representation of where they come from (epistemic claim).
2. I imagine hearing non-located sounds (sounds which in fact are not located in space) (metaphysical claim).

Phenomenology justifies at best the first claim, and further argument is needed to deduce the second claim. One cannot directly infer, from the fact that we can perceptually represent a sound without representing its location, that we can perceptually represent a non-located sound.

5. O'Shaughnessy's View

Strawson's contention is that we can *imagine* non-spatial auditory experiences. This is compatible with the claim that at least *some* auditory experiences represent sounds as being located in space (as he himself acknowledges in the quoted passage). Other philosophers have gone further and rejected even this claim. O'Shaughnessy writes:

Thus, hearing a sound to be coming from point p is not a case of hearing it to be *at p*. This is because the sound that I hear is *where I am* when I hear it. Yet this latter fact is liable to elude us because, while we have the auditory experience of hearing that a sound *comes from p*, we do not have any experience that it is here where it now sounds. (Rather, we work that one out.) And this is so for a very interesting reason: namely, that we absolutely never immediately perceive sounds to be *at* any place. (O'Shaughnessy 2000: 446; emphasis in original)

Here, O'Shaughnessy makes a positive and a negative claim. The positive claim is that we normally hear sounds to be *coming* from a particular place. The negative claim is that we do not hear sounds to *be* at any place.

In our terminology, O'Shaughnessy endorses a *proximal* theory of sounds. First, sounds are metaphysically dependent on the hearer. As such, their location can only be that of the hearer herself: 'the sound that I hear is *where I am* when I hear it'. Second, an error theory of auditory perception is avoided because the spatial content of audition is explained by the fact that we hear *sources* (and not sounds) to be located in space. (Surely we can hear moving objects. If, as some have claimed, events cannot move and sounds are events, the objects we hear as moving cannot be sounds; they must be space-occupying, material objects.) Matthew Nudds (Chapter 4) endorses O'Shaughnessy's point, suggesting that we explain away the putative examples in which one is tempted to say that one hears sounds as located as really cases in which one hears the location of the sources of the sounds.

Let's take stock. Strawson defends an a-spatial theory of both sounds and auditory perception. O'Shaughnessy favors an a-spatial theory of auditory perception (with respect to sounds), but rejects the non-spatiality of sounds.

Strikingly, both Strawson's and O'Shaughnessy's arguments against spatial theories of auditory perception concern the *egocentric* spatial content of auditory experience, namely the kind of content which enables the hearer to locate entities relative to her (to the left, to the right, above, below, in front, or behind). However, auditory perception can have various forms of *non-egocentric* spatial contents. These must be taken into account in fully evaluating the prospects of a-spatial theories of auditory perception.

In what follows, we shall discuss two forms of non-egocentric spatial content. Auditory perception can be said to have non-egocentric spatial content insofar as it represents (i) material sources of sounds as having spatial properties, and (ii) sources and sounds as being spatially distinct entities.

6. Hearing Sources

Auditory perception can give one access to the spatial structure of the world even if it does not have an egocentric spatial content. One can auditorily perceive the constituting *matter* as well as the *internal structure* of sources, whether or not one is able to locate them relative to oneself (think again of the underwater perception; you may not be able to tell where it is relative to you, but you know there is an approaching *boat engine*). One can know just by hearing that an object is hollow, or that it is composed of several interconnected parts. (Think of shaking a closed box containing various tiny objects.) In such cases, one perceives with one's ears various ways in which material objects take up space.

There is a venerable view according to which the primary object of auditory perception is always the sound, and we perceive at best indirectly its source as a space-occupying entity. For instance, Berkeley wrote in his *Three Dialogues between Hylas and Philonous* that 'when I hear a coach drive along the streets, immediately I perceive only the sound, but from experience I have had that such a sound is connected with a coach, I am said to hear the coach' (1948–57: 204). On this view, the perception of the source would be necessarily *epistemic* in Dretske's (1981) sense. We can hear the sound produced by a car, but we cannot hear the car itself. At best, we can hear *that* the car is humming.

As an alternative to the venerable view, consider the hypothesis that perception of sounds is always perception of dynamic states of affairs involving sounds *and* sources. On this hypothesis, sources are as much primary objects of perception as sounds themselves.

Two remarks about the alternative view are in order. First, the view does not entail that one is always able to *recognize* the source on the basis of hearing the sounds it produces. Perhaps one can only think of the source in *demonstrative* terms, such as 'that noisy thing'. Abstract electronic music is an interesting case in point. When listening to this kind of music, one may have no idea of what is producing the sounds one hears—except perhaps loud-speakers.

Second, the 'sources' that we hear are not always concrete, mesoscopic objects.[4] When we hear thunder, for instance, we do not perceive any such object. Still, our auditory perception is about a mass of material molecules involved in a complex vibrating event which either is or constitutes the sound we hear.

Nudds (Chapter 4) acknowledges that auditory perception represents material sources and some of their (static and dynamic) properties. He observes that auditory perception has a *dual* content: one hears material objects *as* the sources of the sounds one also hears. Nevertheless, Nudds refrains from identifying these sounds with the events happening to or within the sounding objects: he

---

[4] Meaning objects that are cohesive, bounded, three-dimensional, and move as a whole. For references and a critical discussion of the role of objects in cognition, see Casati (2003).

writes, 'the sounds we experience normally correspond to their sources—to the things that produced them' (p. 75). The sound one hears when a bell is struck is distinct from the vibrating event happening to the bell. We do not wish to go into the details of Nudds's argument here, but let us make two observations relevant to the evaluation of the Located Event Theory.

First, the Located Event Theory is about the personal level of auditory experience. As such, it is of course compatible with the sub-personal-level claim that information about the object can be extracted directly from properties of the *sound wave* by a process that involves auditory grouping, no part of which requires the auditory system to represent how the *object* is actually vibrating.

Second, Nudds makes much of the possibility that our experience of sounds is veridical even when those sounds do not correspond to their sources. This possibility does not entail that sounds can in fact be detached from their sources. Again, when hearing sounds produced by loudspeakers, one has a perfectly veridical experience as far as the most basic content of experience is concerned—one may just not hear these sounds *as* being produced by loudspeakers. The illusion discussed by Nudds, in which one seems to hear a single sound that is in fact produced by two sources, is more complex. Suffice it to say that it is consistent with the Located Event Theory to claim that in such a case, one's auditory experience is at least partly veridical: one hears audible events located in a more or less definite direction. One has a veridical auditory experience to the effect that something is happening there. After all, one is *surprised* when one finds out which objects are causing our auditory experience, and how. One is indeed fooled about the number of objects and events involved, but one's auditory experience still correctly represents that something audible was going on around there.

7. Hearing Distinctively

The ability to perceive distinct or segregated sounds and sound sequences is known as 'auditory streaming'. Possession of this ability is essential to following a conversation involving several people. It is also involved in the so-called 'cocktail party effect' (Cherry 1953)—the sudden capture of one's attention by a familiar noise in an auditorily clogged environment.

There are pathological cases in which patients lack this ability because their brain cannot process relevant spatial information about sounds. We can describe these cases using the distinction between a 'what' system and a 'where' system underlying auditory experience at the sub-personal level.

The what/where distinction is borrowed from vision science. As is now well known, psychophysical, functional, and anatomical considerations suggest that visual information from the visual cortex follows two distinct pathways: a ventral pathway and a dorsal pathway. The ventral pathway carries information about *what* object there is to be seen, whereas the dorsal pathway carries information about *where* the object is in egocentric space. One of the arguments in favor of this divided architecture is the existence of *double dissociations*: in pathological cases, there can be what without where (*visual disorientation*) and where without what (*visual agnosia*).[5]

In the auditory case, similar double dissociations have been documented. In pathological cases, there can be *what* without *where* (*spatial deafness*) and *where* without *what* (*auditory agnosia*). In the latter case, the patient is unable to recognize sounds but can locate them in egocentric space. In the former case, the patient recognizes sounds (acoustic and semantic recognition is preserved) but is unable to locate them in egocentric space. Spatial deafness should be distinguished from 'deaf hearing', which is the analogue of the phenomenon of 'blindsight' in the visual case. Patients with spatial deafness, unlike those with deaf hearing, have conscious experiences of sounds.[6]

Interestingly, patients with spatial deafness often complain about noisy environments, because for them sounds tend to merge in a cacophony. This suggests that spatial deafness is accompanied by a failure to segregate sounds.

However, this is an oversimplification. There are rare cases of spatial deafness in which subjects have a preserved ability to hear distinct sounds. The patient studied by Bellmann and Clarke (2003) is unable to localize stationary sounds in egocentric space, and to hear sound sources as moving. Still, unlike other spatially deaf patients, she has a preserved ability to segregate sounds and sound sequences.

One experiment which reveals this exploits the 'masking phenomenon' according to which a soft sound (for instance, a tawny owl) disappears if a louder sound of similar frequency range (for instance, a helicopter sound) is presented simultaneously. Of course, the more the sounds are spatially separated, the more clearly they are perceived as two distinct sounds. In the experiment, the owl sound is presented as a target in front of the subject and the masking, and the helicopter sound is presented at various positions relative to it. The subject, who cannot perceive the apparent motion of the masking sound, is instructed to test whether the target is present or not. When the

---

[5] See Ungerleider and Mishkin (1982), Goodale and Milner (1992), and Jeannerod and Jacob (2003). The relevant distinction is not always interpreted as a distinction between 'what' and 'where', especially in the case of vision.

[6] See Clarke *et al.* (2002).

two sounds have different virtual locations, she reports hearing two spatially superimposed sounds, and when they have the same location, she reports hearing just the masking sound.

Bellmann and Clarke (2003) hypothesize that the patient implicitly uses *spatial* cues (in the case in point, inter-aural time differences) to segregate simultaneous sound sources. Thus spatial processing would have at least two functions: localization (the 'where' system) and facilitation of perception (performed as part of the 'what' system).

We can try to imagine what it is like for this patient to have auditory experiences. She is unable to locate the sounds she hears relative to each other or relative to her. As far as the egocentric content of her experience is concerned, her situation is somewhat analogous to that (described in an earlier section) in which a normal subject hears sounds underwater. However, perhaps in contrast to the latter situation, she is able to hear several *distinct* sounds simultaneously. Contrary to other patients with spatial deafness, she does not seem to be bothered by noisy environments (she works in a crowded supermarket).[7]

The foregoing discussion suggests that the ability to perceive *distinct* sounds or sound sequences is at bottom a *spatial* ability. This is relevant to the interpretation of Strawson-like thought-experiments. When we imagine having an auditory experience devoid of egocentric spatial content, we can still imagine hearing distinct sounds simultaneously. For instance, we do not want Strawson's Master Sound (the accompanying sound to every experience in Strawson's thought-experiment) to mask the particular sounds to be heard at the 'places' it embodies. However, given the spatial nature of auditory streaming, it follows that the sounds themselves that are imagined are spatial, or spatially linked to material sound sources, whether they are represented as such or not.

But is auditory experience of distinctness entirely devoid of spatial content? Consider again the sounds to be heard underwater. Their experienced spatial location is not definite. This can mean either of two things. The first possibility is that sounds are heard as completely *non-spatial*. The second is that sounds are heard as being *somewhere*, but that there is no specific location at which they are heard as located.

Now consider the case in which a subject non-egocentrically but distinctively hears two simultaneous sounds. Our suggestion is that the subject's auditory experience has a minimal spatial content: she hears two sources as being spatially separate. Of course, such content is *topological* rather than metrical; she cannot

---

[7] There are here intriguing questions which, as far as we know, are not addressed in the relevant publications, for instance: Can the patient enjoy music? Can she hear multiple sounds as chords?

tell, on the basis of her auditory experience, *how far* the sources are located relative to themselves and relative to her. Still, she can tell that they are not at overlapping, hence connected, locations—she has some spatial information about material objects producing sounds in the spatial world. Hearing distinct sounds and sources is hearing them as having distinct spatial locations, where of course what counts as distinct locations depends on the hearer's auditory acuity.

8. Conclusions

Egocentric spatial content is not the only form of spatial content auditory perception can have. One can hear the material sources of sounds occupying space in many different ways, and one can hear sounds and sources as having distinct positions in space, even if one does not perceive what spatial relation one bears to these objects. (In fact, we may speculate that construing all spatial content as being ultimately egocentric is a prejudice of the post-Kantian, phenomenological tradition.)

What is the relevance of these claims to the evaluation of a-spatial theories of sounds and auditory perception? Well, first of all, the possibility of imagining having (or, for that matter, just having) an auditory experience devoid of spatial egocentric content does not show that such experience is intrinsically non-spatial. In fact, it is quite difficult if not impossible to imagine having a recognizably auditory experience which has no spatial significance whatsoever. At the very least, a convincing a-spatial theory of auditory perception is still forthcoming.

Another point concerns a-spatial theories of sounds. Let's suppose, for the sake of argument, that sounds do not have spatial locations. Can one then maintain the claim that auditory perception is intrinsically spatial, at least in the non-egocentric ways we have tried to highlight above? More precisely, can one avoid an error theory of auditory perception? The only way to avoid postulating massive error would be to argue that the spatial content of auditory perception concerns material sources rather than sounds. After all, in the relevant cases, sources are what we hear as occupying space or as being distinct objects. We do not think that this is a promising way. When one hears sources as occupying space, one perceives that something happens to and within them, and we see no reason not to identify (at least aspects of) these happenings with sounds. Similarly, when we hear two distinct sources, we also hear two distinct sounds. Sounds *and* sources are heard as having some spatial separation. Once again, we do not see any mystery here: sounds are physical events happening to sources, and the spatial locations of the former depend on the spatial locations of the latter. The Located Event Theory still appears to be the best explanation of these phenomena.

It follows that a-spatial theories of sounds are committed to an error theory of auditory perception even if egocentric spatial content is ignored. Sounds must be located somewhere if auditory perception is intrinsically spatial. Note that our initial appeal to phenomenology is still valid: sometimes (if not most of the time), one hears sounds to be located at their sources. There is no reason to think that egocentric spatial content is less veridical than non-egocentric spatial content. (In general, egocentric spatial content is not essentially different from non-egocentric spatial content. Egocentric spatial content locates the objects of perception with respect to the perceiver, which is of course necessary if perceptual experience is to play a role in orienting action, but the same kind of spatial relations are represented in both cases.) Unless one shoulders an error theory of auditory perception, we maintain that phenomenology favors distal theories over proximal and medial theories of sounds.

References

Bellmann Thiran, A., and Clarke, S. (2003). 'Preserved Use of Spatial Cues for Sound Segregation in a Case of Spatial Deafness'. *Neuropsychologia*, 41: 1254–61.

Berkeley, G. (1948–57). *Three Dialogues between Hylas and Philonous*, in A. A. Luce and T. E. Jessop (eds.), *The Works of George Berkeley Bishop of Cloyne*. Bibliotheca Britannica Philosophica. London: Thomas Nelson and Sons Ltd., vol. 2.

Casati, R. (2003). 'Representational Advantages'. *Proceedings of the Aristotelian Society*, 103 (3) 281–8.

—— and Dokic, J. (1994). *La Philosophie du Son*. Nîmes: Chambon.

—— —— (2005). 'Sounds', in E. N. Zalta (ed.), *The Stanford Encyclopedia of Philosophy* (Fall 2005 edition), available at <http://plato.stanford.edu/archives/fall2005/entries/sounds/>.

Cherry, E. C. (1953). 'Some Experiments on the Recognition of Speech with One and Two Ears'. *Journal of the Acoustical Society of America*, 25: 975–9.

Clarke, S., Bellmann Thiran, A., Maeder, P., Adriani, M., Vernet, O., Regli, L., Cuisenaire, O., and Thiran, J. P. (2002). 'What and Where in Human Audition: Selective Deficits Following Focal Hemispheric Lesions'. *Experimental Brain Research*, 147: 8–15.

Dretske, F. (1981). *Seeing and Knowing*. London: Routledge.

Evans, G. (1986). 'Molyneux's Question', in A. Phillips, (ed.), *Gareth Evans, Collected Papers*. Oxford: Oxford University Press.

Goodale, M. A., and Milner, A. D. (1992). 'Separate Visual Pathways for Perception and Action'. *Trends in Neuroscience*, 15: 20–5.
Hobbes, T. (1651/1839). *Leviathan,* in W. Molesworth (ed.), *The English Works,* vol. 3. London: John Bohm.
Jacob, P. and Jeannerod, M. (2003). *Ways of Seeing. The Scope and Limits of Visual Cognition.* Oxford: Oxford University Press.
O'Callaghan, C. (2002). 'Sounds'. Ph.D. dissertation. Princeton University.
—— (2007). *Sounds: A Philosophical Theory.* Oxford: Oxford University Press.
O'Shaughnessy, B. (2000). *Consciousness and the World.* Oxford: Clarendon Press.
Strawson, P. F. (1959). *Individuals: An Essay in Descriptive Metaphysics.* London: Methuen.
Ungerleider, L. and Mishkin, M. (1982). 'Two Cortical Visual Systems', In D. J. Ingle, M. A. Goodale, and R. J. W. Mansfield (eds.), *Analysis of Visual Behaviour.* Cambridge, Mass.: MIT Press, 549–86.

6

The Location of a Perceived Sound

BRIAN O'SHAUGHNESSY

1. Introduction: A Theory

Looking across a room at a piano, hearing a sound that we describe as 'coming from the piano', one can easily find oneself believing that the sound heard is situated there at the piano. We may think of the sound as an invisible but audible real phenomenal entity that takes up a sector of time as long as it lasts (say, the duration of a quaver or a minim), and a sector of space (located around the site of its cause), and we may think of these two quantitative values as constituting its unique spatio-temporal site. Then we naturally think of such an entity as the object of the auditory perceptual experience; we readily suppose that we auditorially perceive this spatio-temporally circumscribed phenomenal entity at a (perceptible) distance in space and (imperceptible, slight) distance in time. I shall now spell out this theory in a little more detail.

Thus, if the place and time of the impact of a finger on a keyboard is $p_1 t_1$, and the place and time of our hearing the resultant sound is $p_2 t_2$, one is inclined to say that at $p_2 t_2$ one hears a sound occurring 'over there' at the piano at p_1. A model might be the associated visual situation: at $p_2 t_2$ I see a finger movement occurring at t_1 at the place p_1 of the impact which caused the sound. And a more extreme example of the same would be: at $p_2 t_2$ on Earth I see through a telescope a spurt of fire happening on the sun's surface, which took place eight minutes ago (at t_1). In sum, in both the auditory and visual cases one is said to perceive across space and time a phenomenon happening at a specific distance both of space and time. The phenomenon in question is supposed to have its unique spatio-temporal site, a sort of spatio-temporal individuating container. A distinct event of perception then occurs which is said to take place at a different spatio-temporal site, and that perceptual experience takes the aforementioned phenomenal entity as its object. Finally, the perceptual experience is said to register the distance and place in space of its object, which is experienced as taking place 'over there' at p_1. In either case we are said to be concerned with an example of *perception at a distance* in space.

2. Perception at a Distance

(1) I think it is beyond dispute that the above is a correct characterization of the visual situation. It is not just material objects and their visible qualities and relations that we see at a remove in space; we also see at a distance events involving these objects, as well as events which do not. For example, I see at a distance a chair, its color and shape, its proximity to a table, and in addition the event in which it topples over as a result of a push, as well as a purely energic event like a flash of lightning. All such phenomena are seen at a distance; indeed, it is simply inconceivable that they be visually perceived otherwise than at a distance in space. This is not to say one must visually perceive what that distance is, or even the distance itself. One can imagine seeing a red monochrome circle on a monochrome blue ground; being unable to visually distinguish this situation from one in which, with eyes closed, one sees after-imagery; unable also to see the distance between oneself and the red entity, or between the red entity and the blue ground. Yet the red entity might turn out to be a red spherical material object situated at a distance in physical space both from oneself and from the blue ground behind it. Then despite the aforementioned shortcomings in the experience, such an experience would undoubtedly count as an example of seeing at a distance in space. For the truth is that the visual perception of all purely physical non-psychological objects is invariably perception at a distance in space (with the sole exception—at least, as I would claim—of visual perception of the visual mediator, light).

But if 'perception at a distance in space' does not mean perception of the intervening space or its measure, what does it mean? Consider two visual experiences in which the perception of depth is absent: the seeing of an after-image and the seeing of a 'floater' in the eye. Now in my opinion we visually perceive visual sensations like after-images, yet this surely cannot be accounted an example of perception at a distance. By contrast, the perception of 'floaters' is a true example of perception at a distance, even though we perceive neither a specific measure of space nor indeed a space of any kind separating the 'floater' from us, the viewer. What rule lies behind these and other characterizations? It cannot be that the object must be located at a distance from the perceiver, as the example of the 'floater' makes clear. Nor can it be that we see a space intervening between us and the object, as the same example also demonstrates. Rather, it lies in the following property of visual perceptions.

(2) Visual perception is a directional phenomenon, and it takes place from a point in space. This is evident in the familiar phenomenon of perspective, that point I shall call 'the origin-point O'. It is a real point in space, not a mere theoretical construct from the nature of visual experience, and may be presumed to be situated somewhere behind the surface of the eyes. After all, we are unable to see events behind the head or within the optic nerves, but capable of seeing both 'floaters' and lights within the eyes. Thus, a distance exists between the point O and the physical item that is visible to one, a distance which is in principle perceptible. I believe that it is to this feature of visual perception that we refer in speaking of 'perception at a distance'. Such perception at a distance, being directional in character, carries the implication that the perceived object points its near side towards the viewer and averts its far side, a feature which generally but not invariably necessitates the perception uniquely of that near side. (An obvious exception is the seeing of a tree at night through the seeing merely of its silhouette.) In sum, perception at a distance is perception that is such that a distance exists between the point from which we perceive and the perceived object.

(3) A somewhat pedantic objection to this analysis is that one might in a mirror see the very point in one's anatomy from which one sees. This is certainly true, and clearly it would be a genuine seeing of that site, and presumably also a genuine example of 'seeing at a distance'. How can that be? In my opinion this is like asking the question, 'How can one leave home for a walk and finish up at one's front door?' In this situation a succession of visible spaces are brought to view that are such as to lead the gaze back to the origin-point P. Why not? Thus, an object intervening between one's eyes and the mirror—a pencil, for example, placed mid-way between the eye and the mirror so that a space exists between the origin P and the pencil—would be visible. Meanwhile, a further space exists between the pencil and the mirror. And both of these spaces are visible. Finally, a visible space exists between the surface of the mirror and one's retina at P. However, these several spaces double back upon themselves, and this fact is not perceived. This is because we are in the presence here of visual illusions—not of objects, nor of spaces, but of directions—and this feature of the situation negates the force of the counter-example. It implies that one does not perceive a distance between P and P, even though one perceives P from P and sees real distances in so doing. Thanks to illusions of direction, an ultimately null journey is undertaken by

the gaze via a sequence of real but convoluted visual journeys on its part. And yet it is a true seeing of *P*. After all, the place *P* that is near to or in the eye is, in this example of visual perception from *P*, given as situated in a part of a material object in such a manner that the object points one side of itself to the viewer. This teaches us that there can be no such thing as a spatially unmediated seeing of a material object. If no distance exists between *P* and a visible object, then that object must have been given to awareness through the mediation of intervening visible spaces.

So what is the analysis of 'perception at a distance in space'? I think the original analysis remains unscathed. We see objects at a distance in the sense that there exists a space between the origin-point of sight and the perceived object. To which I add the rider: this space is in principle perceptible, and it is such that the directionally given perceptual object points the near side of itself towards the viewer and conceals the remote side.

(4) I am of the opinion that perception at a distance is uniquely visual in type. All other varieties of perception encounter their object without spatial mediation. However, I have no reason for believing this to be a necessity. I can think of no reason why there should not exist a non-visual sense which manages to realize many of the properties of sight, including whatever it is that ensures that visual perception is perception at a distance. But as things are, perception at a distance is invariably visual perception. I believe, and will in what follows attempt to prove, that this holds at least for the case of sound.

3. Light and Visibility

(1) Before I address this central question, I shall digress briefly into the example of light, since I believe the perception of sound finds a close parallel in the case of light. Now it seems to me that a strong case exists for the view that for most of our waking lives our visual fields are filled with light, light which is situated at the retina, and which in the visual experience is conceptually individuated under the heading 'brightness of some degree of some visible item' rather than explicitly as 'light' (see O'Shaughnessy 1984–5, 2000: ch. 16). Thus, if I notice the brightness of an object in the visual field, indeed notice under any heading whatsoever any object which is of *some* degree of brightness, then I have, I suggest, noticed the light at that sector of the visual field; that is, I have noticed what is given merely directionally and two-dimensionally in noticing what is given at a perceptible distance in three-dimensional space.

Meanwhile, to those who doubt that our visual fields are filled with light, I would ask two questions. The first is, 'What is it that violet light has that ultra-violet light lacks, if not visibility? Is it merely a corresponding color?' The second question is, 'Are we to believe that light is simply invisible (like X-rays and gamma rays), that no one has ever set eyes upon sunlight or moonlight?' It seems to me that only theoretical commitments could lead one to entertain such beliefs. But theories ought to be responsible to facts, not the reverse. Then if it is agreed that we sometimes see light, what distinguishes the allowed or agreed cases of light perception from the common or garden examples of seeing in which it is supposedly invisible? I can discover no satisfactory answer to this question.

(2) There are good reasons for believing that the seeing of non-psychological, purely physical objects and events is all but universally at a distance in space. The one exception to this rule being the seeing of the visual mediator: light. It is not just that light does not present one face selectively to us, whatever such a 'face' might be; there simply is no such thing as seeing a space separating us from the light that we see. This is because the light that we see is situated at the retina, which is to say at the very point at which the visual system receives its two-dimensional, causally efficacious input from the environment. A common but natural mistake is to suppose that light is visible only when it takes the form, say, of a searchlight; indeed, it is doubly a mistake, since what we see on such occasions is not a cylinder of light. In reality, the seeing of a searchlight is no more an example of seeing light than is seeing a table—and no less. When we see a searchlight, we see a collective of bright material particles, which is in turn one and the same event as the seeing (in two dimensions) of light from those particles (in three-dimensional material object terms). These are of course contentious claims which I cannot hope to justify here, but I am obliged to mention them, because light seems to me to constitute the one exception to a rule which otherwise holds universally.

(3) Let us apply these contentious claims to the visual perception of material objects in interstellar space. We shall suppose that a vast, orange, stellar object O, subtending the same angle here on Earth as the moon, is situated a light-year away, and is almost always shrouded in complete darkness. Assume that at 6.00 a.m. GMT on January 1, 2004, a burst of light illuminates its surface for one second, and suppose that at 6.00 a.m. GMT on January 1, 2005, we see the object O, which is visible to us for one second. During 2004, a light beam of length 186,000 traveled outwards from O in many directions, including towards the Earth. When it appears in our sky, anyone on Earth might quite

naturally ask, 'What was that light in the sky?' Thus, it seems difficult to deny that at that moment one saw orange light in the sky. Then where precisely was the light that we saw? I suggest that it was here where we are—as well as at the many other places constituting the expanding zone occupied at 6.00 a.m. January 1, 2005, by the light reflected by the stellar object O. More exactly, the sector of the light beam that we saw must have been situated on our retinas. For where else? Not an inch in front of the eyes; for that light has not yet reached the visual system—not mid-way to O, for the same reason, and not a light-year away at O, for the same reason—and, in any case, at 6.00 a.m. on January 1, 2005, the object O happens to be shrouded in darkness.

But suppose it is claimed instead that what really happened was that here on Earth on January 1, 2005, we saw happening on stellar object O a flash or influx of light that took place on O a year ago. That is, we saw a phenomenon of light that occurred on O on January 1, 2004, through seeing light that reached our eyes on January 1, 2005. Now an account of this type would be correct if what I had seen was, say, the movement of a rock on O happening at January 1, 2004. Rather, what is here being claimed is that on Earth on January 1, 2005, I saw light happening on the distant object O on January 1, 2004, through seeing light happening here on Earth on January 1, 2005. This is tantamount to offering a light-representational account of the seeing of light; that is, of the seeing of light that is supposedly set at a distance. The position which I shall attempt to defend is, first, that light is never seen at a distance, and, second, *a fortiori* that light-representationalism is false in the case of seeing light, and, third, that light-representationalism is true of the seeing of material objects—of their qualities and relations, indeed of all visible phenomena other than light.

The facts of the situation seem to me to be as follows. We see a movement of a rock that occurred on January 1, 2004; we see the object O as it was on January 1, 2004; we see the event in which the stellar object O lights up on January 1, 2004; and in each case do so through seeing light that is here on Earth on January 1, 2005. And I make these claims because the analysis of the expression 'see an object light up' is exactly paralleled by the analysis of 'see an object brighten'. Seeing an object light up is not catching sight of a material object *and* visually spatially individuating something shining at a distance out from the object's surface like a spear. It is, I suggest, seeing a brightening of a surface *and* seeing something bright that emanated out from the object's surface. But the latter is not given in three visible dimensions of space, nor is it given *in addition* to the former. It is one and the same event, the light being presented to us in two-dimensional terms. It is the event of at once seeing the two-dimensionally given light *and* the brightness of a three-dimensionally given object. Though the brightening of a surface *is not* an emission of light from its surface, the seeing of one *is* the seeing of the other. This is what a light-representationalist account of the seeing of objects involves. And it is inapplicable in the case of the seeing of light itself, for the simple reason that even though visible light starts up at a distance and we see that light, and even though we see thereby much that lies at a distance—indeed, see the light present here with us in the mode of seeing events at a distance (for example, 'see the brightening of O')—the light that we see is never at a distance. It is on the peripheries of the visual system, probably on or around the retina.

4. Why Sounds are Not Perceived at a Distance

The Nature of Sound: Double-Duration and Mutating Space

(1) My main concern in this chapter is to defend the claim that the perception of sound is never perception at a distance (noting, as we proceed, that almost all that I have claimed of light is true of sound). In other words, my claim is that the model of visually perceiving the movement of a finger over there at the piano keyboard, or for that matter the visual perception of a movement on the stellar object O or on the surface of the sun, is inapplicable in the case of sound. Before I continue the discussion, I think that we should for the moment agree to the claim that sound is identical with shock waves in a medium, though I think that the discussion would proceed along pretty much the same lines were we to opt for a more subjectivist account.

I return to the example of hearing a sound coming from a piano. Then there can be no doubt that the sound in question began its life at the piano at place p_1 and time t_1, nor that at p_2 and almost instantaneously at t_2 it was heard. Moreover, it was at t_2 heard to be coming from p_1. Now in my opinion much confusion on this topic derives from one salient feature of sounds: their essential spatio-temporal dynamism. By this I mean not merely the fact that sounds are transient in the way events are transient, but also that they have a double-duration together with constantly mutating spatial properties. Thus, a sound may be a quaver long—which is a time measure—and yet endure for far longer than that time, while simultaneously and concordant with this feature the sound will possess the property of traveling outward from its source and expanding into space as it does. These properties are implicit in the fact that sound is of the nature of a disturbance in a medium, a phenomenon which is not unlike momentum, in that it is something which can be passed or transmitted from one physical item to another. It is not that we arbitrarily choose to conceptualize sound in these terms. Rather, the numerical identity of the cause of the continuous disturbance phenomena across differing times and spaces seems to force it upon us.

If sounds did not travel through space, they would uniquely have the duration that we hear them to have at any one spatial site, and would in this respect be like a movement or change of color. (However, we would never manage to hear such 'sounds'.) Meanwhile, a sound does not possess a double-location as it does a double-time. Even though sound moves, and thus occupies differing and expanding regions of space as time passes, it does not have some first quotum of space which is a constant in quantity as it moves through space, unlike the time measure of a quaver. While it is true that when the quaver-long sound occurs, a determinate region of space has been occupied by the sound, nonetheless, as the sound travels through space, the region it occupies changes in both identity and extent. Thus, time and space are here disanalogous.

The conclusion I would draw from these few observations is that sound is essentially 'on the move' in both time and space. It is not because it is an event that this is so, since much that 'eventuates' is confined in a unique spatio-temporal envelope. It derives rather from the fact that physical force is something which passes from one bearer to another. Momentum is unlike a movement or change of color, in that it cannot be embedded in a single spatio-temporal site. It is along some such lines that I would explain the essential spatio-temporal dynamism of sound.

(2) Sounds move through space. And they do so even if a subjectivist analysis is correct for sound—along the lines one might offer for smell. Think how one might speak of a 'stink' from the Chemistry lab traveling slowly down the corridors of the science building. And yet, 'stinks' need minds as cooperative partners if they are to so much as exist, as do properties like 'uncomfortableness' in a chair. The uncomfortableness of the chair is relative to certain beings, and so, I suggest, is the 'stink'. Nonetheless, 'stinks' move through space—as do sounds. In short, sounds of their very nature travel, whatever one's theory of sound.

If one could manage to see a sound occurring, say, in a visible gas, one would see a phenomenon expanding into space with determinate boundaries. The identity conditions of this shock-phenomenon are such that we reidentify that phenomenon across different regions of space and time. Meanwhile, there are many other phenomena which cannot be reidentified in this manner, and which, as a result, cannot be said to journey through space and time. Take the case of a sudden spurt of fire on the sun's surface, a movement which we see looking through a telescope in California. Whereas the light from that event traveled through space to California, the movement itself did not budge from the sun.

The Argument against Sound Perception being Perception at a Distance

(1) By contrast with the movement of fire on the sun's surface, the 'clack' of a bat on a cricket ball, occasioned by a stroke which I saw through binoculars almost half a mile away, traveled across the intervening landscape to the place where I was standing, and took a few seconds to reach me. In doing so, the 'clack' moved on from the place of its origin, leaving in its wake an entirely different auditory situation. By the time it had reached me and was heard by me at t_2 ($= t_1 + 2.5$ seconds), the only sounds occurring in the immediate vicinity of the batsman might have been an exclamation or silence. This difference between sounds and phenomena such as movement is surely of the utmost importance. It is highly relevant to the problem under consideration, concerning the location of a perceived sound. It leads naturally to the following argument.

We are supposing that at time t_2 and place p_2 half a mile away from the cricket match, I hear a 'clack' sound s', which was caused by the impact of bat on ball. Meanwhile, at p_2t_2 there exist shock waves that were also caused by that same impact. What is the relation between those shock waves and the sound s' heard by me at p_2t_2? Given the assumption that sounds *are* shock waves, the relation is surely one of identity. We also know that shock waves were created at p_1t_1, and *ex hypothesi* were identical with a sound s occurring there and then at p_1t_1. Then what can the relation be between sound s' and sound s? Surely, also, identity. If it is true that sounds travel through space, one and the same thing must begin existence at some p_xt_x and later at some p_yt_y be the same as whatever was at p_xt_x. In short, in this present case it seems certain that $s' = s$.

(2) What conclusions are we to draw from this account of the situation? Well, if the sound I now hear at p_2 was earlier located at p_1, then the sound which I now hear at p_2 must be located where I am at p_2 and no longer at p_1. In that case, I cannot be perceiving any sound at a distance in space. While it is true that what I now hear *was* at p_1, what I now hear *is* at p_2. And that constitutes a first simple proof of the claim that sounds are not perceived at a distance.

5. A First Argument Pro: Sounds are Perceived at a Distance

(1) Consider a protagonist (Mr X) who holds that the sound one perceives is situated at the place of its origin p_1—in other words, who affirms what is the main and contentious proposition under discussion in this chapter. What account would Mr X give of the hearing at p_2t_2 of the 'clack' of bat upon ball? Whatever answer he gives to this question should be consistent with the following indubitable facts. First, that there are sound waves at p_2t_2 which were caused by the impact. Second, that a sound s' is audible at p_2t_2. Third, that sound s' is heard by a subject at p_2t_2. Then what can Mr X say of the hearing at p_2t_2 of that 'clack'? His answer must depend upon whether he accepts that sounds move through space and leave their original space in doing so.

Suppose he does accept this proposition. In that case he must say that at p_2 the subject hears a sound that *was* at p_1 through hearing a sound that *is* at p_2. However, this 'through' cannot be such as to say that he hears one sound (at p_1) through hearing another sound (at p_2), for if Mr X accepts that sounds endure and travel, he has no choice but to accept that the sound that is heard at p_2t_2 is one and the same as the sound that *was* at p_1t_1. Then it is clear that on such an account of the situation, this cannot be a case of hearing a sound that is at a distance; after all, the heard sound has moved on to p_2. What Mr X is here claiming is not in dispute.

(2) But suppose on the other hand that Mr X does not accept that sounds move through space and leave their original space in doing so. Suppose he construes the well-known auditory phenomena in the following manner. He might claim that something really does travel with the speed of sound, only it is not sound. Rather, it is the audibility of the sound that does the traveling, being borne upon the back of the moving sound waves. This would be to construe the auditory situation as parallel to the visual. Thus, the movement of fire on the surface of the sun 93 million miles away does not itself move through space, whereas its visibility does precisely that, and with the speed of light, being borne upon the back of the light waves coming to us here on Earth.

Let us examine this second theory propounded by Mr X, since it brings us to the main topic of the chapter. As we have noted above, such a theory must take account of the following facts: that sound waves exist at p_2t_2, that they constitute a sound s' at p_2t_2, and that the sound s' is heard at p_2t_2. Then if sounds do not travel but audibility does, and if, in addition, as Mr X claims, we hear sounds that are at a distance in space, this must according to him be a case in which one hears one remote sound (s) through hearing a distinct other local sound (s'). In effect, this is to offer a sound-representationalist account of sound perception. This parallels a light-representationalist theory of light perception, and, likewise, is false. Just as it would be false to say, 'I saw the light on stellar object O through seeing light here on Earth', (but correct to say, 'I saw the object O brighten through seeing light here on Earth'), so it is false to say, 'I heard the "clack" at p_1 through hearing a "clack" here at p_2' (but correct to say 'I heard the bat strike the ball at p_1 through hearing sound here at p_2'). So, at least, I shall argue.

The sound-representationalist theory of sound perception has to cope with a formidable difficulty. Let me express the problem in terms of the parallel phenomenon, light. If it were the case that I saw light that was on stellar object O on January 1, 2004, through seeing light here and now on Earth on January 1, 2005, what property singles out the light at January 1, 2004, from, say, light mid-way on January 7, 2004, which suffices to make the former but not the latter the object of the visual experience? In fact, of all the multiple instantiations of this beam of light between January 1, 2004, and January 1, 2005, what singles out the former as against January 7, 2004, or January 10, 2004? The fact that light struck object O on January 1, 2004, seems in no way to constitute an answer, bearing in mind that the reflected light pre-existed January 1, 2004. I can think of no satisfactory answer to these questions.

Another way to put the same difficulty is this. What is the causal link between the light on O on January 1, 2004, and the visual experience on January 1, 2005, which constitutes justification for saying that the visual experience on January 1, 2005, was the seeing of that light on January 1, 2004, rather than the seeing of light at some other time or place along that beam? The causal intermediary between the light on O and the visual experience here on Earth is the beam of light. The relevant link is not between some phenomenon like the movement of a material object on stellar object O with light on January 1, 2005, here on Earth. Rather, the beam links the occurrence of the very same light energy on January 1, 2004, on O, with light here on January 1, 2005, on Earth. That is, it relates one phase of the phenomenon with an earlier phase of itself! But all the other limitless number of phases are equally engaged, causally. Can phases of the single phenomenon play a justificatory role in the reported perceptual exchanges?

(3) At this point, a brief aside before I resume the examination of Mr X's argument. What, according to me, is the real spatial site of the sound that

the subject heard at p_2t_2? Note that this question must be distinguished from, 'What *was* the site at t_1 of the sound that he heard later at t_2?' The sound has a history, during which time its location mutates from an initial site with p_1 at its center. This region then expands in the form of a thick spherical shell until, say, time t_2, and beyond to other times like time t_3, diminishing all the while in intensity but perhaps in no other respects. Accordingly, looking back later, if we ask the question, 'What was the spatial site of the sound which was caused at p_1t_1?', the reply has to be time-sensitive. We will give one reply if the time is t_1, and another if the time is t_2, and another if t_3, of the kind indicated above. While if we ask the separate question, 'What at t_2 was the site of the sound that the subject heard at p_2t_2?', then the answer will be that it was at p_2, amongst other places.

6. A Second Argument Pro: Sounds are Perceived at a Distance

(1) Despite the implausibility of the representational theory offered by our protagonist Mr X, it has to be recognized that the claim that the perceived sound is at a distance has a strong intuitive appeal, which I shall now try to demonstrate.

One might argue for the view that sounds are perceived at a distance in the following way. First, one could point out that the situation exemplified by the delayed 'clack' is foreign to ordinary experience, and therefore remote from the situations that must have helped to shape our concepts of sound and hearing. If sound waves travel at 1,000 feet per second, then the vast majority of the sounds we hear must be heard within 0.1 second of their onset—in other words, at a point at which human acuity gives out. Therefore, it might be claimed, we should abandon the assumption that either sounds or audibility move through space and time. For example, we should abandon the assumption that if a sound were to happen at some p_1t_1, and to be heard nearby at p_2, then that hearing must have occurred at the slightly different time t_2. Rather, we should say that cause, sound, and hearing all occurred at t_1. In short, one might argue that the movement of sound is no part of everyday experience, categorize examples like the delayed 'clack' as atypical and of merely scientific interest, and take the auditory object of hearing to be a phenomenal entity embedded in a unique spatio-temporal site at some distance from the hearer. In this way, one would at one stroke marshal an argument for the view that sound perception is perception at a distance, and simultaneously block the very strong argument that if sounds move from source to hearer, the sound heard by the hearer must be situated at the hearer rather than at the source, and so cannot constitute a case of perception at a distance.

A further consideration which supports the claim that the perceived sound is at a distance is this: We begin by noting that the only spatial site singled out in the auditory experience is p_1—not p_2, not some spatial zone which includes p_1, and not an expanding zone, but p_1 alone, which happens also to be the place of its cause. Thus, the sound is heard to 'come from p_1', and not heard to 'come from' or to 'be at' any place other than p_1. When the sound occurs, the *mind* goes to p_1, and so does the *perceptual attention*; after all, when the sound occurs, one might very naturally look around sharply at p_1 and nowhere else. Does not this show that one's sense of hearing has discerned something audible at p_1, and that the mind follows that lead? While it may be true that evolution determined that mind and attention go to the site of the sound's *cause*, given that the place of the cause rather than of the sound is what is of real import, since the prime use of sound is to inform us of sound-*makers*, this is in no way inconsistent with the fact that the auditory spatial data uniquely single out site p_1. Moreover, hearing that the sound comes from p_1 entails that the sound or part of it is or was at p_1. All of these facts seem to support Mr X's main thesis. Then are we to be deflected from such significant considerations, first, by the existence of a highly atypical phenomenon—the delay attending distant sounds like the 'clack' of bat on ball—and, second, by the linguistic fact that we speak of hearing the sound 'come from' rather than 'be at' p_1? Why, in any case, should we not interpret this 'comes from' as singling out the place of the sound's cause? That is, as synonymous with 'emanates from' or 'has its origin in'?

(2) These considerations are persuasive. But there is a serious difficulty facing this line of thought, which I shall now address.

The first question I would ask is, 'How does a sound manage on some occasions to be co-present with a perception of itself?' The obvious answer is, 'Through the impinging on the ear of shock waves emanating from the site of the sound's cause'. Then how, according to the above account, are we to understand the causal situation? Is it that the cause C at p_1 produces the sound at p_1, which produces shock waves at p_1, which travel on to us at p_2? Or, is it rather that cause C produces shock waves at p_1, which produce the sound at p_1? Or might it rather be the case that the cause C produces at p_1 a phenomenon which is at one and the same time the sound at p_1 *and* the shock waves at p_1? And are there any other alternatives to these three accounts of the causal situation?

It seems clear that no one is going to claim that the sound at p_1 produced the shock waves at p_1. Equally improbable, it seems to me, is the idea that the relation between shock waves and sound is causal. Meanwhile, the third and presumably obligatory alternative must be wholly unacceptable to those who hold that sound perception is the perception of sound situated at a distance. This is because the sound waves (i.e., the sound) arrive at p_2 and are heard, whereupon those who endorse the theory that the sound at p_1 is perceived at a distance at p_2 find themselves saddled with all the problems facing the representational theory of sound perception, some of which have already been rehearsed in Section 5(2).

In short, despite the intuitive appeal of the reformulated theory expounded here in Section 6(1), there seem to be insuperable difficulties when it comes to spelling out the causal perceptual situation this theory implies.

(3) It remains for me to account for the intuitive appeal of the claim that the perceived sound is at the place of its cause. One has to admit that in some sense the sound does seem to be at the place of its cause. For example, one hears a knock on the door, and it can easily seem to one that the knock is an audible entity situated over there by the door. But now suppose that one were to close one's eyes, and suppose also that one had been set down in some grossly unfamiliar, constantly shifting scene—where then would the sound appear to be? Now make the sound an utterly unfamiliar noise. What is left of spatial perception? As it seems to me, direction and little else. Then do not these considerations strongly suggest that one's *mind* sets the knock at the door, but that one's *perceptual attention* sets it, or more exactly sets it and its origin-point, merely in a particular direction (and perhaps a little more)? True, if we whittled away in comparable fashion at a visible scene, we would rapidly drain away all monocular depth-perception, and yet we really do see the distance separating us from objects. Nevertheless, the auditory and visual cases differ. There is a real distinction to be drawn between actually seeing a specific measure of depth, say the distance separating one from an airplane on a tarmac, and mentally importing but not actually seeing such a measure—for example, the distance separating one from an airplane against a completely homogeneous blue sky, or from the moon. While it is true that in each of the latter two examples one would see depth, it is nonetheless also the case that our minds make use of knowledge which enlarges the spatial content of the experience: we see the object *as* situated at a specific distance which we do not actually perceive, and this 'as' has extra-visual cognitive determinants.

I suggest that this important distinction has application in almost all examples of auditory spatial perception in which a specific site is given.

It seems to me that hearing detects the direction, and to a limited degree also the site of the *origin* of a sound, and that the auditory data are augmented by mental factors leading to one's hearing the sound to be *coming from* a specific site like the door. One's attention goes down a line, and does so to an extent that is to a degree inexact, whereas one's mind tends to complete the journey with exactitude. However, even if these considerations manage to explain one's hearing that the sound *comes from* the door (specifically), why also (and contradictorily) should one hear the sound to be *at* the door? The evolutionary rationale behind the sense of hearing determines that our minds should be concerned with the place of the sound-maker, knowledge of which must have important survival value. This fact goes some of the way to explaining why we tend to experience the sound as at its source, since what matters is getting one's mind and attention uniquely to that site. The ultimate purpose of perception is to lead the mind beyond itself, maximizing the available data as it does, outwards onto the objective and often significant material realities in the environment. Perhaps it is along such lines that we may hope to explain the counter-intuitiveness of what I take to be the true account of auditory space perception.

References

O'Shaughnessy, B. (1984–5). 'Seeing the Light'. *Proceedings of the Aristotelian Society*, 85: 193–218.
—— (2000). *Consciousness and the World*. Oxford: Oxford University Press.

7

Hearing Silence: The Perception and Introspection of Absences[1]

ROY SORENSEN

In the course of demarcating the senses, Aristotle defined sound in *De Anima* as the proper object of hearing: 'sight has color, hearing sound, and taste flavor' (II.6, 418b13). Sound cannot be seen, tasted, smelled, or felt. And nothing other than sound can be directly heard. (Objects are heard indirectly by virtue of the sounds they produce.) All subsequent commentators agree, often characterizing the principle as an analytic truth. For instance Geoffrey Warnock (1983: 36) says 'sound' is the tautological accusative of the verb 'hear'.

I shall argue there is a single exception. We hear silence, which is the absence of sounds. Silence cannot be seen, tasted, smelled, or felt. Only heard.

1. How Hearing Silence Differs from Not Hearing

Hearing silence is successful perception of an absence of sound. It is not a failure to hear sound. A deaf man cannot hear silence.

A parallel comparison holds for seeing darkness (Sorensen 2008: ch. 13). A blind man cannot see the darkness of a cave. His sighted companion can. Darkness *conforms* to Aristotle's principle that color is the proper object of sight. Aristotle (correctly) regarded black as a color. Indeed, he thought the chromatic colors were derived from the achromatic colors of black and white. Contemporary color scientists treat blackness as the appropriate color response to the absence of light (Hurvich 1981: 61).

Is hearing silence just a matter of inferring an absence of sound from one's failure to hear? No, a wounded soldier who wonders whether he has gone

[1] I thank Robert Pasnau and Casey O'Callaghan for correspondence on the location of sounds. I am also grateful to John Kulvicki and Walter Sinnott-Armstrong for comments and discussion of earlier drafts.

deaf can hear silence while being neutral about whether he is hearing silence. He hopes he is hearing silence but neither believes nor disbelieves that he is hearing silence.

Some eccentrics make the wrong inference from silence. They think that when they appear to hear silence, they are actually hearing the music of the spheres. They believe this sound is always present. The eccentrics' belief that there is no silence does not prevent them from hearing silence. In the terminology refined by Fred Dretske (1969: 88), people can 'non-epistemically' hear silence. Hearing in the epistemic sense requires belief. Hearing in the non-epistemic sense is compatible with belief but is also compatible with the absence of belief about what is being perceived. That is why we can be surprised by what we hear and even the very fact that we are hearing it. On August 27, 1883, people on the island of Rodriques heard the explosion of Krakatoa 4,800 kilometers away. They only later believed that the sound was that of Krakatoa exploding. The grammatical mark of the epistemic sense is the propositional attitude construction 'hears that p'. The radar operator hears that a sonic boom is approaching by hearing a beep from equipment monitoring an incoming jet. He hears the beep, not the boom.

Hearing silence does not depend on reflective awareness of the silence. Sometimes we become aware of a lengthy silence only after it has been broken. A marginal kind of sensitivity suffices for hearing silence. Turning off a radio awakens listeners who have fallen into a dreamless sleep.

One may dream a silence that is unreal. Even a match between a real moment of silence and dreamed silence is not sufficient for hearing that silence. There can be veridical hallucinations of silence. Consider a man who experiences auditory hallucinations as he drifts off to sleep. He 'hears' his mother call out his name, then wait for a response, then call again. The cycle of calls and silence repeats eerily. As it turns out, his mother has unexpectedly paid a late night visit and is indeed calling out in a manner that coincidentally matches the spooky hallucination. The hallucinator is not hearing the calls and silence of his mother.

2. Hearing Silence Differs from Detecting Silence by Ear

Like other animals, human beings evolved to detect sounds *and* to detect the absence of sound. However, detection of silence is not enough for hearing silence.

You can *detect* the electric charge of a nine-volt battery with your tongue. But an electric eel can *sense* the electric charge (Keeley 1999). The eel has an organ dedicated to this form of energy.

Human beings have ears that are dedicated to sounds. The nose is devoted to odors, the tongue to flavors. Ordinary people spontaneously demarcate the senses by their corresponding sense organs.

Is using your ears to detect a sound sufficient for hearing it? No, when my neighbor blows a dog whistle, I infer by ear (from the puffing noise) that there is an ultrasonic sound. But I do not hear the ultrasonic sound.

Are your ears necessary for hearing sound? Hearing aids show that people hear with defective ears—thanks to mechanical assistance. As the substitute components improve, the entire ear may become dispensable. Technological progress suggests that, in principle, ears are not needed to hear.

This extrapolation carries over to hearing silence. The hearer of silence can rely on a hearing aid or a stethoscope. When an unconscious pilot leaves his microphone on, flight controllers can hear silence in the cockpit.

The perception of silence must be direct in other respects. For instance, one cannot hear silence by listening to a remote sound meter that sounds an alarm when there is silence at a monitored location. You can hear *that* it is silent by means of a sound. But you cannot hear silence by means of a sound. There is a difference between hearing silence and hearing the effects of silence.

3. Silence Does Not Sound Like Anything

Are you *seeing* this sentence or are you *hearing* it? You can easily answer without checking whether you are using your eyes rather than your ears. You can tell by introspection. This suggests a second strategy for demarcating the senses: appeal to the characteristic experiences of the senses (Smith 1990: 239).

The appeal to characteristic experiences explains why future people may see and hear with prostheses. It also handles the perception of darkness. There is a color experience dedicated to darkness. Black is commensurable with other colors. For instance, black (visually) resembles purple more than pink.

Sense datum theorists were naturally attracted to this mode of demarcating the senses. In the original edition of *Perception* owned by Dartmouth College, H. H. Price (1933: 39) proceeds to its logical conclusion: 'We are never destitute of tactual data; and very rarely (if at all) of auditory ones, for what we call "silence" can be heard'. 'What?' has been scribbled in the margin. The reader's puzzlement was not quieted by Price's footnote: 'When I say, "There was silence" I mean something like "My auditory data were of faint intensity and none of them differed greatly from any other"'. The reader balked because silence does not *auditorily* resemble any sound. Silence has no loudness, timbre, or pitch.

Westerners are amused by *shiiin*, the Japanese onomatopoeia for silence. But English has *shush, hush,* and *shhhhhhh*. Perhaps we are imitating white noise in lieu of a sound of silence.

Silence does *acoustically* resemble white noise; there is an absence of discriminable tones. When you turn on a hotel fan to mask the conversation in the next room, you mix sounds. Just as white light combines all of the frequencies of visible light, white noise combines all the frequencies of sound. Hearing white noise is hearing all sounds; hearing silence is hearing no sound.

White noise varies in loudness. Silence is invariant. White noise sounds different from silence. When you eventually halt the hotel fan, you are relieved by the silence. You can hear the difference between white noise and silence.

The experience of silence has a qualitative aspect. A hypothetical scenario featuring an acoustic scientist Audrey can bring this out. She lives in a noisy environment and so has never experienced silence. Audrey knows the physical aspects of silence. But she wants to experience silence and so constructs a soundproof chamber. When she enters the chamber, Audrey learns something; what it is like to hear silence. She closes her eyes to listen more intently. She enjoys the silence as others might enjoy the burble of a brook or the jug-o-rum of bullfrogs.

Audrey is introspecting an absence of auditory sensations while perceiving an absence of sound. A patient with an ear problem can introspect gaps in his auditory sensation of a rising tone. Audrey wanted more than the gap in her sensations. She wanted an auditory gap that originates through healthy hearing of an external state of soundlessness.

Austen Clark prefaces *A Theory of Sentience* (2000) with a thought experiment. Suppose your senses are discreetly incapacitated, step by step, so as not to disturb your meditation on an abstract issue (which you are conducting with closed eyes to avoid the distraction of the senses). At the end of the subtractions, you are conscious but sense nothing. Clark draws the lesson that consciousness without sentience is conceivable. But his thought experiment can be diverted to make a different point. When you become aware that you are blind, deaf, and generally senseless, you are *introspecting* an absence of sensations. For you no longer perceive anything. Introspection is your only remaining means of detecting the absence.

Perhaps we approximate Clark's scenario during sleep. If there is a stage of sleep in which there are no sensations, we might introspect this absence. The introspection of absences is continuous in Clark's scenario. But in a 'dead sleep', introspection also stops. The gap in consciousness could still be sensed after the fact. This awareness may be what we are appealing to when we compare death to sleep. If experience requires sensations, then one cannot know what death is like (at least from a first-person perspective). However, if there is a qualitative aspect of experiencing absences, then we do have an inkling of what it is like to be dead. Our experience with intermittent losses of consciousness gives us a basis to extrapolate to the permanent loss of consciousness.

It does seem overly modest to say that a man knows nothing more of death than a toddler or a turtle or a termite. The man has a better understanding of what it is like to be dead because he has more experience of mental gaps and better means to learn from those experiences.

Parmenides characterized death as the brother of sleep. A man's thought is a ratio of light and night in his body. With sleep as in death (and aging), that ratio changes in the direction of night.

For according as the hot or the cold predominates, the understanding varies, that being better and purer which derives from the hot... But that he [Parmenides] also attributes sensation to the opposite element in its own right is clear from his saying that a dead man will not perceive light and heat and sound because of the loss of fire, but that he will perceive cold and silence and the other opposites. And in general, all being has some share of thought.   (Robinson 1968: 124)

I have three disagreements with Parmenides' development of the analogy between sleep and death. First, the dead do not continue to have experience (even at a low order). Second, the perception of the dark, the cold, and silence is perception of what is not rather than what is. (Absences are not at the low end of a hierarchy of being.) Third, Parmenides fails to privilege silence. Hearing silence is the most negative of perceptions: there is nothing positive being sensed *and* no positive sensation representing that absence.

Clark's scenario has a haunting resemblance to death. The introspected absence of sensation is global. Audrey's scenario is the more typical introspection of a local absence of sensation.

Audrey might want to share her chamber's silence with her husband. Audrey's husband could get the same type of experience from another soundproof chamber. But his attachment to Audrey makes him want to hear the silence she arranged and to jointly experience the same particular silence she is hearing.

Audrey's silence differs from the silence of others because it is caused differently. The darkness of caves varies the same way.

When you see the darkness of a cave, you can introspect the visual sensation of darkness. When you feel the cold (which is the absence of heat) there is a different sensation to introspect. The qualitative aspect of the cold sensation explains our surprise at the burning sensation of dry ice.

Introspection may help us correct confusions between absences. Many people believe that it is darkest before the dawn. But they are actually experiencing the extremity of another privation. The landscape cools off all night making it *coldest* just before dawn. (It is darkest at midnight when the sun is farthest from sunrise and sunset.)

Synesthetes have experiences that trespass between sense modalities. They 'see' sounds and 'hear' colors. Possibly there is synthesia for the perception of absences. A synesthete who can 'see' coldness might have a special impetus for asserting, 'It is darkest before the dawn'.

When you hear the silence of Audrey's chamber, there is no sensation of silence to introspect. You instead introspect the absence of auditory sensations. Just as you can perceive the blanks between the words of this sentence, you can introspect gaps between sensations. Just as blanks can sometimes be organized into a gestalt pattern, gaps can form patterns that can be introspected holistically.

Pauses are used to chunk speech into perceptual units. But pauses themselves can be unified within sounds. Musicians exploit these higher order forms of silence perception.

In addition to having neurons that fire in recognition of tones, we have neurons that fire in recognition of pauses and gaps in tone sequences (Hughes *et al.* 2001). Perhaps this is the neurological basis for the introspection of missing sensations.

One can say pauses sound like something in the guarded way one can say that blanks look like something. You can show someone a blank by pointing at one on a page. You can exhibit a pause by having the learner listen to a specimen. The difference between seeing a blank and hearing a pause is psychological. We are primed to see letters and so see the absence of letters. We are primed to hear sounds and so hear the absence of sounds. Absences are relative. They draw their identity from their relata.

The differences between absences are nonetheless objective. Members of the British National Antarctic Expedition were killed in 1912 by the cold. They were not killed by silence or the dark.

Empiricists trace all of our knowledge of the world through the senses. Sensationalists further say that sensations are the basic elements of experience. But there is no auditory sensation of silence.

Sensationalists overestimated the role of introspection. We hear mostly without introspection. That includes hearing silence. There may be creatures that hear silence despite their total inability to introspect. Audrey can savor silence because she can attend to the workings of her own mind.

4. Silence Can Last Indefinitely

Pauses depend on sounds just as the hole of a doughnut depends on the doughnut. If the sound does not return, then the pause does not last indefinitely.

Sounds are generally short-lived and this makes pauses even briefer. There are exceptions. J. O. Urmson claims, 'the sound of Niagara Falls outdates our most cherished antiquities' (1968: 119).

One complication is that the roar of Niagara Falls has not been continuous. On March 30, 1848, the flow was stopped for more than 24 hours by an ice jam upstream. Those who believe that sounds cannot survive interruption will date the present roar to no earlier than 1848. Surely there have been longer pauses in the 12,000-year history of Niagara Falls. Silence can have an impressive duration.

Indeed, there is no upper bound on how long silence can last. Imagine Seshat, the Egyptian goddess of mathematics, is counting one number per second. She utters the prime numbers out loud but silently counts the composite numbers. Thus there are moments of silence in Seshat's oral recitation:

$$2, 3, \ldots, 5, \ldots, 7, \ldots, \ldots, \ldots, 11, \ldots, 13, \ldots, \ldots, \ldots, 17, \ldots$$

There are infinitely many prime numbers but they become sparser and sparser down the number line. Thus the stretches of silence become longer and longer without limit.

Can silence be infinitely long? First, let us consider whether a *sound* can be infinitely long. Aristotle believes that infinity is always potential, never actual. So he would reject an example of Apollo continuously playing his lyre. At any point in the future, the immortal Apollo is finitely old and so his music is always finitely long.

However, Aristotle appears to have believed that species have an infinite past. So consider the murmur of a hive of bees. The murmur is the collective effect of many bees. None of these sources is essential; a bee can leave while the murmur continues. Indeed, the murmur can continue through the gradual replacement of all the bees. Since we are imagining the species has an infinite past, the murmur could be infinitely old while each bee is only finitely old.

The murmur of the innumerable bees illuminates the metaphysics of sound. If sounds are dependent on a particular source (O'Callaghan 2007: ch. 5), then no sound is older than its source. Yet the murmur is older than any bee.

Given the Big Bang theory of the origin of the universe, no sound is infinitely old. It could still be the case that there are sounds older than any source. And there could be silences older than any sound. A pause can take place between distinct sounds.

5. Silence is a Proper Object of Hearing

The third strategy of demarcating the senses is by what they sense. Common sensibles (number, shape, magnitude, motion, and so on) are available to more than one sense. Proper objects (flavor, odor, sound, color, tactile qualities) can only be directly accessed by a single sense.

Contemporary defenders of this demarcation strategy face two problems. First, proper objects seem like an obsolete, arbitrary grouping. They fail to mesh with the natural kinds that have come to light through modern physics and chemistry.

A second classic problem is specifying the proper objects for sight and touch. Theorists have trouble coping with the sheer variety of what we see and feel (Sanford 1976).

In contrast, the specification of the proper object for hearing seems straightforward. That is why proponents of the proper object strategy, such as George Berkeley, model sight on sound. They are drawn more strongly to sound than to odor and flavor because sound better approximates the spatiality and informativeness of what we see.

However, silence shows that the proper object of hearing is, in one respect, trickier to specify than the object of vision. There is a color corresponding to the privation of light. But there is no sound corresponding to the privation of sound. Silence presents a new anomaly for those who wish to demarcate the senses by their proper objects.

6. Odorlessness is Not a Proper Object of Smell

We detect that a rhododendron flower is odorless by smelling it. But do we smell its odorlessness? We detect that tofu is flavorless by tasting it. But do we taste its flavorlessness?

These perceptions of absences are less clearly sensings than seeing darkness or hearing silence. The reason is that odorlessness and flavorlessness are of only marginal significance to human beings.

Obviously, odors and flavors are important to us. Loss of the sense of smell or taste merits medical attention. But our encounters with odorlessness and flavorlessness are dispassionate. We do not savor the flavorlessness of tofu. We do not stop to smell the odorlessness of the rhododendrons.

There are characteristic emotions (such as disgust) and behaviors (nose wrinkling) associated with odors. Odors also interact with other sense modalities. Much of what is ascribed to taste is actually odor.

The emotional significance of odors is a legacy of our hunter-gatherer past. Odors provide clues to food, sex, and health. Odorlessness did not betoken opportunity or danger. That makes odorlessness emotionally flat.

Contemporary people make use of unprecedented substances that are dangerous because they are odorless. Natural gas must be adulterated with mercaptan (which stinks like sulfur) to make leaks noticeable to homeowners. If odorlessness, in the Paleolithic Era, had been an exploitable sign of danger or opportunity, we would have evolved behavioral responses to odorlessness and emotions that organize those behaviors. As it stands, odorlessness scores low on criteria we use to distinguish between sensing and detecting.

Silence scores much higher. There are characteristic behaviors to generate and detect silence. There are also characteristic emotional reactions to silence. Moreover, the perception of silence is integrated with the rest of the perceptual and cognitive systems.

7. Silence Interacts

Primitive predators quiet down to listen. Their prey freezes to avoid making a sound.

Sophisticated animals find silence instructive at a meta-level. Hush is a sign that conspecifics have acquired information. Just as animals stop and orient to an unexpected sound, they stop and orient to an unexpected silence. When a group is wary of predators or other enemies, silence may serve as an alarm.

What begins as a natural sign can develop into a conventional sign. Pauses punctuate conversation, playing a variety of grammatical roles. 'Signs of omission' are easier to see in written language (Sorensen 1999). We can afford to demarcate written words with blanks. Inscriptions last for a while. Spoken words linger only in working memory. To get the message across quickly, the speaker runs his words together. He merges some of the phonetic components of a word into a single sound. Hearers are equipped with a module that unpacks these co-articulated phonemes.

Small silences have a phonetic effect (Dorman *et al.* 1979). For instance, insertion of a gap makes the difference between hearing a sound as 'split' as opposed to 'spit'. I conjecture silence also has intermodal effects. Studies that show how sound affects vision generally feature a baseline condition of silence. I interpret some of these studies, such as the motion bounce illusion (Sekuler *et al.* 1997), as evidence that *silence* affects vision. In the silent condition, two moving dots pass through each other in an 'X' pattern. When a 'ping' sound is inserted at the point of intersection, subjects instead see the dots as bouncing off of each other. The usual interpretation is that the intermodal effect is restricted to the sound condition. My speculation is that the silence condition also features an intermodal effect. The silence encourages us to interpret the dots as shadows.

Our emotional reactions to silence are shaped by what silence signified to our hunter-gatherer ancestors. Silence is a sign of abandonment or ostracism. This may be at the root of our fear of silence. In his *Pensées*, Blaise Pascal writes, 'The eternal silence of these infinite spaces frightens me' (2003: 61). Why be afraid of nothing? Because silence is associated with disapproval and estrangement.

Since silence conveys nothing on its own, it is usually sensitive to context. Depending on the circumstances, silence can convey assent, dissent, or uncertainty. Its message is heavily context-dependent. Silence can be an expression of respect. One of the rituals of Armistice Day is a two-minute silence held at 11 a.m., 'the eleventh hour of the eleventh day of the eleventh month' (the time at which the 1918 armistice went into effect, bringing World War I to a close).

The gesture of silence can be amplified by darkness. In Poland, the death of Pope John Paul II was commemorated on the *evening* of April 8, 2005. The lights were switched off in homes throughout the nation to reinforce five minutes of silence.

Signs for silence are conventional. Egyptian statues represent the child Horus as a naked boy with his finger on his mouth. This incarnates the hieroglyph for 'child'. However, the Greeks and Romans misinterpreted the finger placement as a gesture for silence. Thus was born Harpocrates, the god of Silence and Secrecy.

The meaning of silence is also colored by its physiological effects. Silence is welcome when it betokens the resolution of a crisis: blood pressure ebbs, heart rate declines, muscles relax. Silence is conducive to concentration. Seneca trained himself to philosophize amidst the hubbub of ancient Rome. But most thinkers require a refuge from noise.

8. Silence has a Location

Peter Strawson (1959: 65–6) denies that sounds have an intrinsic location. We can correlate sounds with various locations. For instance, I was taught to calculate my distance from the source of thunder by counting the seconds between the lightning strike and the thunder. I was told that one second equals one mile. (Much later I got the bad news that this rule involves a five-fold underestimate of the proximity of the electrical discharge.) Strawson concedes there are *contingent* connections between sound and locations. He denies there is an auditory field comparable to the visual field. A purely visual concept of space is possible (even if impoverished). A purely auditory concept of space is impossible.

Matthew Nudds suggests that Strawson's point is that we can see a portion of space as empty. When you look at a ring, you are aware of the hole. Your awareness does not depend on seeing anything in the hole.

> It is this visual awareness of places where there is nothing which has no auditory equivalent. We are simply not auditorily aware of empty places—there's no difference between not experiencing a sound at some place, and experiencing no sound there. One may hear nothing at some place, but in doing so one never comes to be aware of a place at which there is no sound—one is simply unaware. (Nudds 2001: 213)

But a teacher can hear the silence of her classroom while also hearing a lawnmower outside. She thinks, 'It is silent in here but noisy out there'. A conductor can hear silence from the left half of the choir while hearing the right half singing.

To develop the import of these counter-examples, I rely on the principle that the location of silence is parasitic on the location of sound. Just as a shadow borrows the shape and volume of a material object that might have filled its space, silence borrows the direction and location of a possible filler sound. This draws me into controversy about the location of sounds.

Most sounds have a location and a direction. Hearing a sound in your right ear and silence in your left ear can help you pinpoint the location of faint noise. This is just a limiting case of exploiting the sound shadow formed by your head. Your head blocks incoming sound waves like an island blocks ocean waves. If the waves are strong enough to make it around the head, there will be an informative time delay and an informative change in amplitude.

The location of a sound does not have the same qualitative status as loudness, pitch, and timbre. We postulate the property of timbre because sounds that have the same pitch and loudness can sound different. For instance, a cello is mellower than a flute even when the two have the same pitch and loudness. Two sounds from different locations can sound alike (Clark 2000: 60). If identical watches are placed on either side of your ears, the tick-tock of one watch will be heard as coming from the left and the tick-tock of the other watch will be heard as coming from the right. But it will be the same type of sound. The same goes for a gap in the tick-tock. You will simultaneously hear a brief silence on the left and a brief silence on the right.

Just as there is nothing intrinsic to a tone to indicate its location, there is nothing intrinsic to a silence. Yet we can locate silence. After a building collapse, rescuers sometimes only hear silence from the rubble (Clark 2000: 61).

Chapters on sound location are standard fare in textbooks on hearing. Robert Pasnau (1999) says these chapters conflict with the chapters identifying sound with waves in a medium. If sounds are in the medium (typically for human beings, the air), then they are all around us. Sounds would lack the specific, differential locations we commonly attribute to them. The sound waves are actually moving *away* from the source. So Pasnau thinks that those who identify the sound with sound waves must say hearers are overly narrow when locating sounds. The sound waves caused by a loon are not just at the loon; the sound waves are all around you.

To avoid postulating an illusion, Pasnau denies that sounds are sound waves. He argues that sounds are vibrations of the source (or, more cautiously, that sounds supervene on these vibrations). Objects have sounds in the way they have colors. The sound of a tuning fork is more intense as you approach it. But this is no more a change in the sound than a change of its look as you approach it. The tuning fork's image size increases but it does not really look bigger. A tuning fork sounds better in a concert hall than in a meadow in the same way it looks better in daylight than twilight. In the dark, we cannot see the color of the tuning fork but its color still exists. In a vacuum, we cannot hear the sound of the vibrating tuning fork but it still has a sound. The conditions are just bad for hearing the sound.

Casey O'Callaghan (Chapter 2; 2007) agrees that Pasnau has shown incoherence in the acoustic textbooks and in common sense. However, he tries to preserve insights from the wave theory of sound by characterizing sounds relationally as events in which a source disturbs a medium. Since there is no medium to disturb in a vacuum, O'Callaghan denies that a vibrating tuning fork makes a sound in a vacuum.

Whereas Pasnau thinks each sound is a property of the source (like the redness of a rose), O'Callaghan agrees with the wave theorist that sound is a particular. O'Callaghan believes that sounds depend on their sources; each sound must have a source, is always located at its source, and can never switch sources. The wave theorist acknowledges that sounds have sources, but grants autonomy to sounds. The wave theorist is impressed by the linearity of sounds. Instead of rebounding, our voices pass right through each other.

Pasnau and O'Callaghan grant that the wave theory of sound is endorsed by both science and common sense. However, they are so impressed by the problem of locating sound that they are willing to take on both of these authorities.

Since the critics of the wave theory concede its positive merits, my defense of the wave theory is restricted to defusing the appearance of locational inconsistency. I will rely on the traditional method of semantic ascent. Instead of immediately discussing the location of sound, I discuss how we answer 'Where is it?' questions.

Many things are located by their edges. But we also locate by centers. American football players tackle a deceptive runner by concentrating on his center of gravity (or a salient spot that tends to project from that location such as the runner's navel). When edges are indiscernible, we have no choice but to use interior features (the eye of a hurricane, the foci of an elliptical orbit, the solar system's center of mass). Authors of acoustics textbooks describe sound as a train of waves emanating from a source in all directions. In a uniform medium, the shape of the sound will therefore be a sphere. Discussants of a sound are talking about a big, rapidly expanding phenomenon that envelops them. Since the edges of the sound are unknown, orientation by the boundary is forbidden by H. P. Grice's (1975) maxim of quality ('say only what is true'). Grice's conversational maxim, 'Be informative', rules out the true but trivial remark that the sound is in the air. So if anything is to be said about the location of sound, it must be in terms of its source. The speed and invisibility of the sound prevent us from experiencing the sound like an approaching water wave. Any attribution of movement to the sound will be based on movements of the source.

The linguistic aspects of locatedness are more pronounced in an example that is free of the phenomenology of locatedness. Seismology textbooks have a chapter that defines an earthquake as a series of shock waves caused by failure of brittle rocks in the Earth's crust. The same textbook will have a chapter explaining how to locate an earthquake by triangulating to the hypocenter of the failure. The hypocenter (also called the focus) is below ground and so not readily recognized by us surface dwellers. So seismologists refer to the point on Earth's surface directly above the hypocenter: the epicenter.

Are the seismologist's assertions jointly consistent? The earthquake cannot be located at just the epicenter *and* just at the hypocenter. If the earthquake is the train of seismic waves emanating from the hypocenter, then the quake is in its medium and so encompasses a wide area. These waves are moving *away* from the hypocenter. One wave front briefly heads *toward* the epicenter but then spreads out from there.

One might try to avoid seismic incompatibilism by claiming that 'earthquake' is ambiguous. Disputes over 'Where is New York located?' are dissolved by noting that 'New York' is ambiguous between Manhattan, the City of New York (which includes Manhattan as a borough), and New York State. However, disputes over 'Where was earthquake?' cannot be dissolved in the same way. For all seismologists agree that there is no sense in which an earthquake is its epicenter. The epicenter is a physically insignificant point. Despite the belief that earthquake is not its epicenter, seismologists truthfully answer 'Where was earthquake?' by locating its epicenter. (An earthquake without an epicenter is geometrically possible. Suppose the hypocenter is at the center of Earth, or, more plausibly, near one of the bulges of Earth—at the poles or along the equator. In these cases, there is no unique point closest to the surface.)

Seismic incompatibilism is less attractive than acoustic incompatibilism. First, we have no seismic phenomenology of locatedness that orients our perception to a central source. Second, the effects of pragmatics are salient for earthquake location. The term 'epicenter' was obviously introduced in the same instrumental spirit as 'arctic circle'. Answers to 'Where was earthquake?' will vary with our purposes. The feeling of inconsistency is being generated by our failure to relativize to these interests.

We need a uniform treatment of the acoustic and seismic cases because their phenomena overlap. If the primary wave of an earthquake refracts out of the rock surface with a frequency of more than 20 hertz, human beings will hear that earthquake as a low rumble. Seismologists search for a mechanism to explain how acoustic sand dunes manage to boom (Nori *et al.* 1997). Blind mole rats communicate by thumping their heads on tunnel walls (Nevo *et al.* 1991). The seismic waves are conducted through their bones and processed by their auditory system. (Deafened mole rats stop thumping.)

If we had a phenomenology of earthquake location, it would resemble our phenomenology of sound location. We would feel earthquake as being at a central point. For the phenomenology of locatedness is governed by perceptual counterparts of Grice's maxims of quality and informativeness. It is true but trivial to say the quake is inside Earth. The edges of the quake were never accessible during the evolution of our perceptual systems. So we would have no choice but to orient toward the center.

Since silence is an absence of a sound, it has a location where the sound would have been. Pragmatic factors will dictate whether we locate in accordance with the absent sound waves or the inactive source.

Since I identify sound with acoustic waves, I think silence is the absence of acoustic waves. Waves are positive phenomena that depend on a medium. Silence is equally dependent but negative.

When silence is a missing sound, it will be at the place normal for sound. Consequently, the location of silence is as predictable as the location of sound. Even so there are surprises. On May 18, 1980, there was silence around the 60-mile blast zone of Mount Saint Helens. Since sound waves travel faster in warmer air, they bend toward cooler air. The volcano's sound was bent up to the higher altitudes. About 15 miles up it was refracted down again. Thus the sound had the shape of an expanding doughnut. The silence was the hole of the doughnut.

The dramatic, deceptive effects of refraction, wind, and temperature move some scientists to speak of sound mirages. In the early twentieth century, physicists interested in foghorn design became alarmed by how sound can skip over areas near the coast and land at amazing distances from the foghorn. More ominous than these 'false alarms' was the inaudibility of the foghorn at the zone targeted for warning. Physicists carefully plotted the boundaries of this dangerous area of silence (Mallock 1914: 73–4).

Are there silence mirages? Given that all mirages work by refraction, silence mirages are impossible because silence cannot refract. Given that all echoes work by reflection, silence cannot echo. However, silence does abide by almost all the laws of refraction and reflection. In earlier work on light, I have called this parasitic behavior 'para-refraction' and 'para-reflection' (Sorensen 2008: chs. 6, 7). Although there cannot be silence mirages or silence echoes, silence may have para-mirages and para-echoes.

Transmission devices enhance our ability to hear silence at a distance. One of Jack Benny's radio skits was designed to underscore his miserliness. A mugger confronts Benny: 'Your money or your life!' After a prolonged pause, members of the audience begin to laugh. They realize that Benny must be thinking it over. To get the joke, the audience must not interpret the absence of noise as a failure of transmission. They must interpret the absence as being conveyed from the radio station.

One cannot hear the difference between silence from a source and silence from an absent source. This indeterminacy is explored in Harold Pinter's 1959 radio play, *A Slight Ache*. The play appears to have three characters: Edward, Flora, and a match-seller invited into their home. Edward and Flora confide much to the match-seller. But he never speaks. Eventually the audience begins to wonder whether Edward and Flora have just invented the match-seller. The play is propelled by the unresolved question of whether the silence is coming from anyone. That is why the play is difficult to televise and very difficult to stage.

The composer Leopold Stokowski is reported to have once reprimanded a noisy audience: 'A painter paints his pictures on canvas. But musicians paint their pictures on silence. We provide the music, and you provide the silence'. The silence of the audience does not mean that the auditorium is silent. The whole point of refraining from making sounds is so that the musicians can fill the hall with music.

9. Silence is Not a Limiting Case of Sound

One might protect the generalization that we directly hear only sounds by characterizing silence as a zero-value sound. There are determinables such as temperature that have zero-value determinates. Temperature is defined as a measure of the average amount of molecular motion. The absence of all motion is not an absence of temperature. It is a temperature of zero degrees on the Kelvin scale. Yuri Balashov (1999) argues that some key physical properties conform better to a zero-value hypothesis than to an absence hypothesis: spin, electric charge, and perhaps mass (to cover photons).

If silence is a zero-value sound, then what does it have zero of? We cannot answer that it has zero decibels because zero decibels is the softest sound audible to average, young human beings. This is about the softest sound that any creature perceives. The vibrational amplitude of the air at zero decibels is only about the diameter of a hydrogen atom. It is counterproductive to make the ear more sensitive than this. Creatures hear by virtue of *systematic* variations in air pressure. Variations below zero decibels become random because of the thermal agitation of molecules (Brownian motion). If the sensitivity were set too low, the hearer would be distracted by a meaningless, ambient hiss.

Some speculate that young children can hear the hiss. If so, they might lose this useless capacity in the way they lose the capacity to discern phonemes that are not used in their language.

But let us suppose that we can make sense of zero loudness in some other way. Does silence have a zero-value pitch? Is silence very low or very high? Can two silences of equal loudness and pitch differ in timbre? Unlike loudness and pitch, timbre cannot be scaled from high to low.

Two sounds can cancel out because of destructive interference. But the interpenetrating sound waves still exist. They are superimposed at the 'dead spot' and will become audible as they move beyond this area. If the destructive interference is perfect, then the hearer will report hearing nothing. Or he might take himself to be hearing silence. But he is actually hearing two sounds that sum to zero. The listener can be shown that he is not hearing silence by deactivating one of the sources of the sound waves. He will then hear one sound that has always been there.

A loud sound differs from silence by its magnitude. Lowering the sound decreases this difference. However, a quiet sound still differs from silence in the way a slightly dirty sheet differs from a clean sheet. To be clean is to have no dirt, so it is not a kind of dirtiness.

10. Challenges to the Possibility of Silence are Misguided

What counts as 'no dirt' depends on the domain of discourse. The knife emerging from your dishwasher is clean but not by surgical standards. What counts as silence in a classroom does not count as silence in a recording studio.

Restricted quantification is common. Silent movies have no sounds from the recorded events but do have accompanying music.

'Silence' is an absolute term like 'flat' and 'certain'. There is a tendency to privilege the strictest standard—to let the quantifiers be 'wide open'. One is then apt to conclude that there is very little silence.

The impossibility of silence is a popular thesis in literary philosophy. Maurice Blanchot (1986: 11) writes, 'Silence is impossible. That is why we desire it.' Georges Bataille (1988: 16) characterizes 'silence' as a 'slipping' word because it is 'the abolition of the sound which the word is; among all words it is the most perverse, or the most poetic: it is the token of its own death'.

The impossibility of silence is most methodically championed by philosophical commentators on music, especially since John Cage's 4'33" (Davies 2003: ch. 1). The first performance was at the Maverick Concert Hall in 1952. The pianist opened and closed the piano at the end of each movement but did nothing else. The performance lasted four minutes and thirty-three seconds. Most of the audience interpreted the performance as a period of silence—or perhaps three periods to correspond to the three movements in Cage's program notes. Of course, there was the usual coughing and shuffling plus noises that wafted in from outside. But the audience did not count these sounds as part of the performance, just as these sounds do not count as part of the performance in the case of conventional music. According to Cage, the original audience

> ... missed the point. There's no such thing as silence. What they thought was silence, because they didn't know how to listen, was full of accidental sounds. You could hear the wind stirring outside during the first movement. During the second, raindrops began pattering the roof, and during the third the people themselves made all kinds of interesting sounds as they talked or walked out. (Kostelanetz 1988: 65)

Cage claims that 4'33" was inspired by a trip to the soundproof anechoic chamber at Harvard University. Although he expected silence, he heard a high noise and low noise. The engineer explained that the high noise was the sound of his nervous system in operation and the low noise was the blood in circulation.

More precisely, the high noise is tinnitus—'ringing in your ears'. Prolonged exposure to loud noise (greater than 90 decibels) accelerates the degradation of hearing associated with aging. Students are rightly alarmed by how much hearing damage is revealed by their visit to an anechoic chamber.

Just as the skeptic lowers the standards of doubt to show that certainty is impossible, Cage lowers the standards of sound to show that silence is impossible. He does this quantitatively by letting very faint sounds count as sounds. He does this qualitatively by letting a wide variety of phenomena count as sound (even the auditory sensations that are not due to sound waves).

'Silence' echoes the semantic unclarities of 'sound'. Is 'sound' the vibration of an object? Or does it consist of the waves produced by the object? Or is it the auditory sensation produced by those waves?

The skeptic is notoriously difficult to refute. Cage is more vulnerable. The sounds he mentions are observer-dependent. An anechoic chamber is silent when unoccupied.

But can silence be heard? Yes, because we can overcome observer effects. A microphone can be installed in the chamber and we can listen from outside. The sounds we make in the listening booth are not in the empty anechoic chamber.

Can silence be *directly* heard? Yes, engineers could drill an ear-sized hole into the anechoic chamber so that only Cage's pinna fits. Cage's head and torso would still be outside the chamber.

In any case, tinnitus and the sound of circulating blood are logically contingent aspects of human observers. There could be less intrusive observers.

Mark Nyman summarizes Cage's project:

> It is a well-known fact that the silences of 4'33" were not, after all, silences, since silence is a state which it is physically impossible to achieve... 4'33" is a demonstration of the non-existence of silence, of the permanent presence of sounds around us, of the

fact that they are worthy of attention, and that for Cage environmental sounds and noises are more useful aesthetically than the sounds produced by the world's musical cultures. 4′33″ is not a negation of music but an affirmation of its omnipresence. (Nyman 1974: 22)

4′33″ fails to prove any of these theses. True, if we set standards high, silence is hard to achieve—as is flatness, straightness, and cleanliness. But there is no reason to privilege high standards. A high standard is appropriate for special purposes (surveying, pharmaceutical production, and so on). But normally a high standard conflicts with our master goal of being informative. Engineers raise standards only when new technology makes it feasible to meet them.

When John Cage sets a high standard for silence, we are naturally inclined to set a high standard for other absolute terms. Cage says that sound is omnipresent. Given unrestricted quantification, 'omnipresent' means *everywhere* in the universe. The vast majority of the universe is empty. And empty space is silent. As Cage grew older, he expressed optimism about the future of music. He took solace in the conviction that there will always be sound. But given high standards, 'always' means *every time*. The laws of thermodynamics doom the universe to heat death. Everything, everywhere, will end in silence.

References

Aristotle. (1987). *De Anima*, in J. L. Ackrill (ed.), *A New Aristotle Reader*. Princeton: Princeton University Press.
Balashov, Y. (1999). 'Zero-Value Physical Quantities'. *Synthese*, 119: 253–86.
Bataille, G. (1988). 'Inner Experience', in L. A. Boldt (trans.), *L'Expérience Intérieure*. Albany, NY: SUNY Press.
Blanchot, M. (1986). 'The Writing of the Disaster', in A. Smock (trans.), *L'Ecriture du Désastre*. Lincoln, NE: University of Nebraska Press.
Clark, A. (2000). *A Theory of Sentience*. Oxford: Oxford University Press.
Davies, S. (2003). *Themes in the Philosophy of Music*. New York: Oxford University Press.
Dorman, M., Raphael, L., and Liberman, A. (1979). 'Some Experiments on the Sound of Silence in Phonetic Perception'. *Journal of the Acoustical Society of America*, 65: 1518–32.
Dretske, F. (1969). *Seeing and Knowing*. Chicago, Ill.: University of Chicago Press.
Grice, H. P. (1975). 'Logic and Conversation', in D. Davidson and G. Harman (eds.), *The Logic of Grammar*. Encino, Calif.: Dickenson, 64–75.
Hughes, H. C., Darcey, T. M., Barkan, H. I., Williamson, P. D., Roberts, D. W., and Aslin, C. H. (2001). 'Responses of Human Auditory Association Cortex to the Omission of an Expected Acoustic Event'. *Neuroimage*, 13: 1073–89.

Hurvich, L. M. (1981). *Color Vision*. Sunderland, Mass.: Sinauer.
Keeley, B. L. (1999). 'Fixing Content and Function in Neurobiological Systems: The Neuroethology of Electroreception'. *Biology and Philosophy*, 14: 395–430.
Kostelanetz, R. (1988). *Conversing with Cage*. New York: Limelight.
Mallock, A. (1914). 'Fog Signals: Areas of Silence and Greatest Range of Sound'. *Proceedings of the Royal Society of London*, series A, 91(623): 71–5.
Nevo, E., Heth, G., and Pratt, H. (1991). 'Seismic Communication in a Blind Subterranean Mammal: A Major Somatosensory Mechanism in Adaptive Evolution Underground'. *Proceedings of the National Academy of Science*, 88(4): 1256–60.
Nori, F., Sholtz, P., and Bretz, M. (1997). 'Booming Sands'. *Scientific American*, 277: 84–9.
Nudds, M. (2001). 'Experiencing the Production of Sounds'. *European Journal of Philosophy*, 9: 210–99.
Nyman, M. (1974). *Experimental Music: Cage and Beyond*. New York: Schirmer Press.
O'Callaghan, C. (2007). *Sounds: A Philosophical Theory*. New York: Oxford University Press.
Pascal, B. (2003). *Pensées*. Translated by W. F. Trotter. New York: Dover.
Pasnau, R. (1999). 'What is Sound?' *Philosophical Quarterly*, 49: 309–24.
Price, H. H. (1933). *Perception*. New York: Robert M. McBride and Co.
Robinson, J. M. (1968). *An Introduction to Early Greek Philosophy*. Boston: Houghton Mifflin Company.
Sanford, D. (1976). 'The Primary Objects of Perception'. *Mind*, 85: 189–208.
Sekuler R., Sekuler A. B., and Lau, R. (1997). 'Sound Alters Visual Motion Perception'. *Nature*, 385: 308.
Smith, A. D. (1990). 'Of Primary and Secondary Qualities'. *Philosophical Review*, 100: 221–54.
Sorensen, R. (1999). 'Blanks: Signs of Omission'. *American Philosophical Quarterly*, 36(4): 309–21.
—— (2008). *Seeing Dark Things*. New York: Oxford University Press.
Strawson, P. (1959). *Individuals*. London: Methuen.
Urmson, J. O. (1968). 'The Objects of the Five Senses'. *Proceedings of the British Academy*, 54: 117–31.
Warnock, G. J. (1983). *Berkeley*. Notre Dame, Ind.: University of Notre Dame.

8

The Sound of Music[1]

ANDY HAMILTON

According to the acousmatic thesis defended by Roger Scruton and others, to hear sounds as music is to divorce them from the source or cause of their production. Non-acousmatic experience involves attending to the worldly cause of the sound; in acousmatic experience, sound is detached from that cause. The acousmatic concept originates with Pythagoras, and was developed in the work of 20th century *musique concrète* composers such as Pierre Schaeffer. The concept yields important insights into the nature of musical experience, but Scruton's version of the acousmatic thesis cannot overcome objections arising from timbral and spatial aspects of music, which seem to relate sounds to the circumstances of their production. These objections arise in part from music's status as a performing art rooted in human gesture and behavior. Hence I defend a *twofold thesis* of 'hearing-in', which parallels Richard Wollheim's concept of 'seeing-in': both acousmatic and non-acousmatic experience are genuinely musical and fundamental aspects of musical experience. Musical sounds are essentially part of the human and material worlds. While the acousmatic thesis is ultimately unpersuasive, however, the concept of the acousmatic places an interesting interpretation on traditional debates. It is also the case that a more developed musical understanding tends towards the acousmatic. I conclude by considering some implications for the metaphysics of sound, arguing that the twofold thesis of the experience of music implies that one can experience the location and production of sounds through hearing alone.

[1] I am grateful for detailed comments on this article from Jason Gaiger, Louise Richardson, Roger Scruton, and Roger Squires; to Abigail Heathcote for translation of the passage in Schaeffer; and again I am especially indebted to David Lloyd for his technical expertise in sound recording and reproduction. Thanks also for comments from Paul Archbold, Philip Clark, Michael Clarke, Christoph Cox, Stephen Davies, Jonathan Harvey, Brian Marley, Mark Rowe, Nick Southgate, and Michael Spitzer, and the audience at the Philosophy and Music Conference in Pretoria 2004. I am indebted to the British Academy for providing funding to attend this conference, and to Chérie Vermaak for organizing it.

> The hills are alive with the sound of music
> With songs they have sung for a thousand years
> The hills fill my heart with the sound of music
> My heart wants to sing every song it hears
>
> (Oscar Hammerstein II)

1. The Acousmatic Experience of Sound

When Oscar Hammerstein II succeeded Lorenz Hart as Richard Rodgers's lyricist, the artistic quality of Rodgers's song output declined, reaching a nadir with *The Sound of Music*. Unlike Hart, Hammerstein made Rodgers compose the songs to fit the lyrics, and Hart would never have inserted redundant words to make the line scan, as he did with 'the sound of music'. The hills could have been alive with the sound of cow-bells, rally cars, log-rolling, or avalanches, but music is the sound itself. Nonetheless, the phrase became the title of the song, musical, and film, and the rest is history (Wilder 1972).[2] But what exactly is musical sound? Or, perhaps an alternative way of asking that question, what does the experience of musical sound involve? Reversing Mr Hammerstein's ill-conceived formulation, there is a persisting tendency within music aesthetics and musical thought in general to say that musical sound is not really the sound *of* anything—at least not anything material. Music, the most abstract of the arts, is divorced from the material world.[3] On this view, while worldly sounds are characterized in terms of their producer, to hear sound as musical is to separate it from its producer. Hence the *acousmatic thesis* explored by this chapter: that to hear sounds as music involves divorcing them from the worldly source or cause of their production.

The concept of the acousmatic was made explicit in the 1940s and 1950s by *musique concrète* composers such as Pierre Schaeffer. *Musique concrète* is early electronic music, which, since it has no performers, has no visual element to engage audiences or listeners. Its exponents believed that 'listening without seeing' allowed sounds to be more easily appreciated for themselves. In contrast, when traditional, non-electronic music is performed, the circumstances of its

---

[2] Conductor Sir Thomas Beecham's delightful quip, that 'The English people may not understand music, but they absolutely love the noise that it makes', trades on the same redundancy (Watson 1991: 331).

[3] This thesis is extended to sound itself by Strawson (1959) and Nudds (2001)—on which see the final section.

production are normally fully visible. The concept of the acousmatic is developed by Roger Scruton (1997) in *The Aesthetics of Music* and in the present volume (Chapter 3), in the form of what I call the 'acousmatic thesis'. He writes that, in musical experience,

[W]e spontaneously detach the sound from the circumstances of its production, and attend to it as it is in itself... The acousmatic experience of sound is precisely what is exploited by the art of music... The history of music illustrates the attempt to find a way of describing, notating, and therefore identifying sounds, without specifying a cause of them.   (Scruton 1997: 2–3)

Later he writes that

The person who listens to sounds, and hears them as music, is not seeking in them for information about their cause, or for clues as to what is happening. On the contrary, he is hearing the sounds *apart* from the material world. They are detached in his perception, and understood in terms of their experienced order... [T]he notes in music float free from their causes... What we understand, in understanding music, is not the material world, but the intentional object: the organization that can be heard *in* the experience.[4]   (Scruton 1997: 221)

Zuckerkandl (1969: 273), quoted by Scruton, in some ways anticipates this general position: 'Tone... does not lead us to the thing, to the cause, to which it owes its existence; it has detached itself from that; it is not a property but an entity'.[5] The key thought behind the acousmatic thesis is this: An economy of meaningful sound appears to liberate sounds from the need to have a worldly source, and so music escapes the gravitational pull of its causal origin. What remains is its non-worldly or musical cause or rationale.

The acousmatic thesis does not yield a criterion that distinguishes music from speech; it seems equally plausible to claim that to experience sounds as meaningful speech involves divorcing them from the source of their

---

[4] In fact, Scruton's commitment to the acousmatic thesis is not totally clear, as he sometimes speaks of it as ideal rather than actualized: 'in day-to-day matters, we leap rapidly in thought from the sound to its cause, and speak quite accurately of hearing the car, just as we speak of seeing it. But the phenomenal distinctness of sounds makes it possible to imagine a situation in which a sound is separated entirely from its cause, and heard acousmatically, as a pure process' (Scruton 1997: 11–12).

[5] The acousmatic thesis is implicit in the work of a number of writers, such as Spitzer (2004), and Edward Lippman: 'Hearing is satisfied with its own objects, and has no need to relate them to further objects and events of the outside world. This is especially evident in the case of tone and tonal configurations. ... [Sonority's] ontological status is clearly that of an object peculiar to hearing; it can not be located at all in environmental space ... [M]any objects of hearing alone—even of binaural hearing—tend spontaneously to be perceived as immanent rather than external, while multisensory objects are normally perceived as external, and special effort is required to apprehend them as immanent.' However, he adds that 'Music abstracted entirely from the environment... is a [Western] creation of the 19th century' (Lippman 1977: 46–7, 50).

production. Music and speech are distinguished by the fact that although both impose a structure on sounds, speech is a cognitive matter while music is essentially the object of aesthetic attention. What makes a sequence of sounds into speech is that they are meaningful, and it is not essential—indeed it may be a distraction or a barrier to understanding—to appreciate the sounds 'as sounds'. With music, in contrast, it is essential to appreciate the sounds as sounds, in the sense that we do not attend to them for the information they yield, whether through their non-natural meaning (as in the case of speech) or natural meaning.[6] (It is not clear how one should take the suggestion, inspired by Murray Schafer's work on soundscape (1969, 1977), that before the Renaissance, speech and music were not divorced.)[7]

In fact, while no one would deny that musical experience involves attending to sounds as sounds, as opposed to attending to sounds to gain information through their non-natural or natural meaning, the claim that experiencing music involves 'hearing sounds as sounds' may be resisted. There is a real dispute here, and many composers and musicians will take a stand against contemporary proponents of 'sound art'. For it might be argued that there is a non-musical sound art which involves hearing sounds as sounds in some distinct non-musical sense. Sound artist Francisco Lopez (2004), for instance, explains how he is 'fighting against a dissipation of pure sound content into conceptual and referential elements ... trying to reach a transcendental level of profound listening that enforces the crude possibilities of the sound matter by itself'. From a very different perspective to that of sound artists, Roger Scruton asserts that in a vital sense musical listening precisely does not involve hearing sounds as sounds. While Jerrold Levinson's (1991) persuasive account holds, with modernists and postmodernists, that no intrinsic properties of sound—melody, rhythm, harmony—are required for something to count as music, Scruton, agreeing that no intrinsic properties of sound are required, argues in opposition to modernist conceptions that melody, rhythm, and harmony are necessary yet non-intrinsic. For him, these qualities are properties not of sound, but

---

[6] Clearly the issue is a complex one. There is speech that is art—drama and poetry—and speech that is not, and one can attend to the actor's or poet's voice aesthetically. But music is essentially an art at least with lowercase 'a', a claim defended in Hamilton (2007a). It is also true that in contrast to music, speech shows no fixed pitches—see, for instance, Lippman (1977: 52).

[7] According to Van Leeuwen, 'In The Middle Ages and the Renaissance the voice was still a musical instrument and music was embedded in every aspect of everyday life, just as many "less developed" cultures had and still have songs for grinding grains, songs for harvesting crops, songs for constructing houses ... But as clergical plainsong, the cries of night-watchmen, and the chanting of the ABC in schools were replaced by reading aloud, speech was divorced from music, and much flattened in the process' (Van Leeuwen 1999: 1).

of tones—that is, pitched sounds—and sound becomes musical tone when organized by pitch, rhythm, melody, and harmony.[8]

Scruton develops an acousmatic account of tones through traditional defining features of pitch, rhythm, melody, and harmony. Timbral and spatial aspects of music, which seem to relate sounds to the circumstances of their production, are interpreted by him in a manner which he believes is compatible with the acousmatic thesis. He makes various distinctions between the acousmatic and non-acousmatic realms, distinguishing acoustical experience of sounds (non-acousmatic) from musical experience of tones (acousmatic); the real causality of sounds from virtual causality between tones; and the sequence of sounds from the movement of tones that we hear in them. Scruton links the acousmatic thesis with the claim that tone is the intentional object of intrinsically metaphorical musical perception. For him, the objective but phenomenal acousmatic realm exhibits a 'virtual causality' between tones, in contrast to the real causality between sound-producers—musical instruments included—and sounds. Virtual causality is found in melody, where we hear not just change, but movement—a rising and falling in pitch, and tension and resolution. It is also found in rhythm, where talk of movement is metaphorical—only the performer's body and limbs, the instrument, and air molecules, literally move—but essential to the experience of music (Scruton 1997: 92). This important and persuasive thesis of the *necessary metaphorical perception* of music is touched on in Section 4 of this chapter, but reserved for fuller discussion elsewhere.[9]

It is essential to recognize that the acousmatic thesis is a claim about how musical sound is *experienced*—viz., without reference to its physical cause—and not about how it is known to be produced. In acousmatic experience, according to the thesis, listeners know that the sound has a physical cause, but it is not that to which they attend. Rather, they attend to the imagined or virtual causality present in the musical foreground and 'heard in' the medium. What motivates the acousmatic thesis, I would argue, is this process of 'hearing-in', which shows interesting parallels and contrasts with the concept of 'seeing-in', the twofold experience of painting and picturing developed by Richard Wollheim (1980). While the latter phenomenon is much discussed, its aural equivalent is almost entirely neglected, and it is a valuable feature of the acousmatic thesis that it corrects that neglect. However, my conclusion is

---

[8] The issues in this paragraph are discussed in Hamilton (2007a, 2007b).

[9] In the case of movement, the metaphorical transference is not just one way. To say, 'The music moves', is a projection of human bodily movement; but the description of the human bodily behavior is a musical one. One does not just think of music as behaving like a human body, but the human body as behaving musically. The issue is pursued in Hamilton (2007b: ch. 5).

that aesthetic experience embraces the non-acousmatic, just as it embraces the non-representational element in painting, and so ultimately I reject the acousmatic thesis. Nonetheless, I show that it yields significant insights into the nature of musical experience, and expresses an important dichotomy in terms of which the experience of music can be understood. I conclude by proposing a twofold account of musical experience involving both acousmatic and non-acousmatic—a duality apparent in other sound arts and sound-design, and in any aesthetic experience of sound.

Scruton's discussion has received surprisingly little attention. In musicological (as opposed to philosophical) circles, this is in part because of his opposition to certain core features of musical modernism, and the account of tonality and atonality which results (Hamilton 1999). However, the acousmatic thesis conforms with an enduring strand of thought about music, which detaches it from the world, making it the most abstract of the arts—a pure 'art of tones'. This strand of thought reached its apogee with the ideology of absolute music in the 19th century. Both Kant and Hanslick subscribed to it, drawing opposite conclusions about the status of music in conformity with the opinion of their time. Scruton's commitment to a pure art of tones is qualified by his humanism—the understanding that the very concepts of music, dance, and human gesture are interlinked. My argument is that this commitment to a pure art of tones is insufficiently qualified—that the acousmatic thesis neglects the importance of the human production of musical sounds, to which appreciation of music makes essential reference, and which therefore limits music's abstract nature. (Note, for instance, Scruton's comment quoted above that to hear sounds as music involves 'hearing [them] *apart* from the material world'.) It also neglects the way that our experience of music relates it not just to the human body and behavior, but also refers to the nature of sound-producers—the instruments—as physical objects, and the physical phenomena of sound-production. It is not true that music is the object only of metaphorical perception, therefore; attending to sounds as part of the human and material worlds is a genuinely musical part of musical experience.[10]

2. Pythagoras and Musique Concrète

I begin my account of the acousmatic thesis by examining its origins in *musique concrète*, and before that Pythagoras. *Musique concrète* composers espoused an

---

[10] Philosophical humanism concerning music is defended in Hamilton (2007b).

acousmatic thesis but not in the form that Scruton presents it. *Musique concrète* used electronic means to extend the composer's resources to non-tonal sounds. (Non-tonal in the sense of 'not based on tones'—as opposed to atonal, 'not based on the tonal system of major and minor keys'.) Unlike traditional composition, it did not depend on performers to interpret or realize a notated score. When Pierre Schaeffer founded the 'school' of *musique concrète* in the 1940s, he used the primitive recording technology of disc-cutters and tape-recorders to create compositions from a montage of everyday and natural sounds—doors slamming, steam trains puffing, and people talking, as well as more traditional musical materials such as the piano and other instruments. These sounds were modified in various ways—played backwards, cut short or extended, subjected to echo-chamber effects, varied in intensity, and with certain frequencies were filtered out or reinforced—which in later *musique concrète* had the effect of obscuring or destroying clues about the source of the sounds (Wishart 1986: 45). The term *concrète* is meant to convey the idea of working directly or concretely with sound material, in contrast to the composer of traditional music who—according to exponents of *musique concrète*—works indirectly or abstractly through a system of notation which represents the sounds to be made concrete. *Concrète* also conveys the genre's concern with natural, real-world source-sounds, though recorded electronic sounds were not forbidden.

While *musique concrète* typically treats 'worldly' sounds from everyday life, pure electronic music—initially known by its German designation *elektronische Musik* and sometimes referred to as sound synthesis—is produced at least in part by computer synthesis. However, despite the ideological divide between *musique concrète* and pure electronic music, most practitioners were not purists. Stockhausen's early *Gesang der Jünglinge*, which uses recordings of a child singing, has a close affinity to *musique concrète*; and in *Hymnen* from 1966–7, he used recordings of national anthems that, although transformed, were recognizable as such. Although the French and German traditions have since merged, some distinct tendencies are still apparent.[11] But the historic divide between *musique concrète* and *elektronische Musik* now centers on how the composition is realized in the performance space as much as the kind of material exploited—hence the contrast between sound *diffusion* of acousmatic music and sound *reproduction* of taped music.[12]

---

[11] On the history of electronic music, see Holmes (2003).
[12] Diffusion of a stereo source over a multi-channel loudspeaker system is the norm in *musique concrète*, and implies live control during performance and interaction with the performance space. In the electronic music tradition, in contrast, each channel on the tape is mapped onto a single

For Schaeffer, a composition is experienced acousmatically when a curtain has been lowered between its constituent sounds and their previous worldly existence. In this situation, which the medium of recording privileges, sounds are treated as objects divorced from their sources or causes. Schaeffer took the term 'acousmatic' from descriptions of Pythagoras' practice of lecturing to students from behind a screen, so that they would attend to the words and not the speaker; the esoteric or religious sect of Pythagoreans were called *akousmatikoi*, 'those willing to hear'. According to Burkert (1972: 192), Timaeus tells of a five-year period of probation, during which the new disciple was obliged to listen in silence, and did not even see Pythagoras; the sage's voice emerged from behind a curtain. Thus Schaeffer wrote: 'We can, without anachronism, return to an ancient tradition which radio and recording follow in the same way today, restoring to hearing alone the entire responsibility of hearing a perception ordinarily leaning on other sensory evidence' (1966: 91).[13] In music, Pythagoras is best known for his discovery of the natural harmonic series, according to which consonant musical sounds are related by simple number ratios. The followers of this shadowy figure split into Acousmatics and Mathematicians, and it was the scientific *mathematikoi* who endured.[14] Correspondingly, two distinct lines of musical influence can be derived from Pythagoras' thought. The first is the regimentation of the natural harmonic series, resulting in tonal music in its broadest sense—music based on tones; that is, determinate pitched sounds of a certain stability and duration.[15] The second way of thinking, not influential until the 20th century, is so-called acousmatic music, which explores the inner nature of sound. In contributing to the latter conception through the development of *musique concrète*, Schaeffer sought an alternative to the tonal music which sprang from the Pythagorean proportions of intervals.

loudspeaker, implying an attempt to replicate the composer's conceived space within the performance space—intervention in performance is concerned solely with balance, not with exploiting the individuality of the space. See Harrison (1999).

[13] My translation.

[14] Unlike other pre-Socratics, there are no statements by Pythagoras which later authorities agree in attributing, and only after Plato is he represented as the head of a philosophical school. Burkert writes that modern controversies over Pythagoras as shaman or philosopher are 'basically nothing more than the continuation of the ancient quarrel between *acusmatici* and *mathematici*... [T]he Platonists... attribute to Pythagoras himself a more sophisticated version of the Pythagorean number theory. Plato and his pupils thus stand in the tradition of the *mathematici*, and it is not surprising that their version carried the day' (Burkert 1972: 197). Burkert himself sides with the *acusmatici*: 'the "wisdom" of a shaman-like "divine man" can stand without the prop of science... Greek science, including Greek mathematics, may well have had another and non-Pythagorean origin' (1972: 208).

[15] Palombini (2004) refers to 'The musical note, a notable assortment of pitch, duration, and intensity, [which] has borne sway over European tradition and laid claim to universality'. This most crucial direction of Pythagorean influence is discussed in Hamilton (2007b: ch. 1).

The *musique concrète* tradition took up the term 'acousmatic', and later exponents often described their work as 'acousmatic music'.[16] They use the term in a much more restricted sense than Roger Scruton, for whom all music is acousmatic.[17] *Musique concrète* composers tend to describe 'acousmatic listening' as 'listening without seeing'—though Schaeffer is concerned not just with how listeners should perceive sounds, but also with the attitude which composers should adopt towards their material. In both cases, he maintains, one should ignore the physical origin of the sounds employed, and appreciate them for their abstract properties. Schaeffer also termed the process 'reduced listening', arguing that recording encourages it both through the possibility of listening without seeing, and of indefinite repetition. Sound reproduction has a double role: 'to retransmit in a certain manner what we used to see or hear directly and to express in a certain manner what we used not to see or hear' (Palombini 2004). In this way Schaeffer seeks to reconcile technology with nature, treating the medium of analogue recording like the curtain which concealed Pythagoras from the *akousmatikoi*—excluding visual experience while at the same time enhancing experience of the sonorous object in the way to which we have now grown accustomed through telephone, tape, and radio.[18]

Strictly speaking, reduced listening should not be equated with listening without seeing; rather, it is listening that is enhanced by not seeing. The object of acousmatic or reduced listening is what Schaeffer, apparently discounting the common sense assumption that sounds are temporal processes rather than things, calls a sound-object (*objet sonore*):

In order to avoid confusing it with its physical cause or with a 'stimulus', it seems that we have based sound objects on our subjectivity. But...far from being subjective, in the sense of personal [and] incommunicable...sound objects...lend themselves quite well to being described and analysed...Such is the suggestion of the acousmatic: deny

---

[16] According to Dhomont (1995), the term 'acousmatic music' was introduced by *musique concrète* composer François Bayle in 1974.

[17] There are different uses of the term even within the rather esoteric domain of electronic music. While 'acousmatic' has particular associations with the variety of recorded electro-acoustic composition known as *musique concrète*, some writers extend it to all electro-acoustic music existing in recorded form and designed purely for loudspeaker-transmission—as opposed to live electronic music in which sounds, originating electronically or from voices and traditional instruments, are generated, triggered, or transformed in the act of performance. Surprisingly, one such writer, quoted in Sadie and Tyrrell (2001) (entry on 'Electro-Acoustic Music' by S. Emmerson and D. Smalley) is François Bayle, head of the Groupe de Recherches Musicales (GRM), leading current practitioners of *musique concrète*.

[18] John Dack reiterates Schaeffer's view, writing that after recording and *musique concrète* or radiophonic transformation, sound 'can now attain the status of a sound object [and] acquires an autonomous identity...' (Dack 1994).

the instrument and cultural conditioning, to put in front of us the sound and its musical possibility.[19]   (Schaeffer 1966: 93)

Also: 'When [sound recognition] is effected without the aid of sight, musical conditioning is shaken up. Often surprised, sometimes uncertain, we discover that much of what we believed was only in fact seen, and explained, by the context' (Schaeffer 1966: 97).[20] Schaeffer recognized that a Pythagorean curtain will not discourage our curiosity about causes, to which we are instinctively drawn. But he maintained that reduced listening counteracts this tendency: 'the repetition of the physical signal, which recording makes possible... by exhausting this curiosity... gradually brings the sonorous object to the fore, [and] progressively reveals to us the richness of this perception' (Cox and Warner 2004: 78). (Compare the repetition of a word—its meaning is forgotten as one concentrates on the sound itself.) At first, where we are ignorant of what is causing the sound, we want to know what it is; with practice, however, the desire dissipates (Chion 1994: 32). It should also be noted that one can desire to *know* the origin of the sound, while at the same time *experiencing* it acousmatically.

To reiterate, Schaeffer's conception of acousmatic music concerns not just how listeners should perceive sounds, but the attitude which composers should adopt towards their material—one which attempts to ignore the physical origin of the sounds they use, and appreciates them for their abstract properties. Schaeffer (1966: 103–28) distinguished four modes of listening ('*les quatres écoutes*'):

(1) Indexical mode of listening (*écouter*): concerned with identification of events responsible for the emission of sound.
(2) Symbolic mode (*comprendre*): sounds as signs, signifiers, or signifieds that are *extra-sonores*.
(3) Naïve reception of a sound's occurrence (*ouïr*): 'I heard something'.
(4) Attention to qualities of the sound itself, without reference to its source or significance (*entendre*).

Both (2) and (4) are acousmatic, but only the latter involves the experiencing of *sounds as sounds* that is characteristic of musical listening—*La recherche musicale*, which Schaeffer proposes is based on a return to 'the sound itself'. The *objet sonore* is an 'in-itself' to be explored by 'bracketing' both significations and causes. Schaeffer (1966: 360–85) aimed to develop the everyday (*banale*) non-referential listening of *entendre* into a specialized semiotic system, equivalent to

---

[19] Translation by Abigail Heathcote.   [20] Translation by Abigail Heathcote.

pre-existing musical and linguistic systems in its relational and abstract nature yet completely different in its development of 'natural' listening.

He constructed this system after beginning his collaboration with Pierre Henry on 'Symphony *pour un homme seul*'. Schaeffer created a syntax for non-tonal as well as tonal sounds, a *solfège* or typology for *objets sonores*—sounds considered in separation from their sources and classified in terms of tessitura, timbre, rhythm, and density (the degree to which the sound-object fills out the sonic spectrum). He had begun developing this semiotic system of reduced listening in his preliminary studies of 1948, in which he recorded percussion instruments, and discovered that any single musical event is characterized not only by the timbre of the main body of the sound, but also by its attack and decay (Manning 1993: 20 ff.).[21] Schaeffer distinguished two elements of the sound-object, the complex spectrum associated with a sharp attack or abrupt change in content, and the more ordered, slowly changing spectrum usually associated with the body of the sound and its decay; the former element can be so disordered as to be a semi-continuous spectrum of noise. In 1952, he produced a definitive syntax in the form of 'Esquisse d'un solfège concrèt', the last chapter of *A la recherche d'un musique concrète*. He consolidated his formidable apparatus in the *Traité des objects musicaux* (1966). It involved three *'plans de référence'*—melodic (the evolution of pitch parameters with respect to time), dynamic (evolution of intensity parameters with respect to time), and harmonic (the reciprocal relationship between parameters of pitch and intensity represented as spectrum analysis). Thirty-three criteria for evaluating the three plans in total were suggested, resulting in around 54,000 possible combinations of sonological characteristics.[22] This syntax attempted to characterize non-tonal sounds independently of their sources.

In invoking the parallel of *solfège* in his taxonomy of musical and non-musical sounds, however, Schaeffer emphasized the connections with traditional musical creation. It seems that he did not consider the possibility that he was creating a category of sound art distinct from music—though that, arguably, was his achievement. Instead, he increasingly felt that he had failed to deprive sounds of their literal connotations, declaring in despair that '*Musique concrète* in its work of assembling sound, produces sound-works, sound-structures, but not music' (Kahn 1999: 110).[23]

---

[21] The implications of this discovery are discussed further in Section 4.
[22] I am indebted to Manning (1993) for this information.
[23] See Hamilton (2007a, 2007b: ch. 2). Indeed, Schaeffer's viewpoint was in many respects conservative. When asked 'What is the exact moment at which something becomes music?', he offered a criterion even more restrictive than Scruton's appeal to tone, referring to the 'traditional testimony...that a

3. A Broader Definition of 'Acousmatic'

The concept of the acousmatic is not restricted to what, for most listeners, is the rather specialized domain of electro-acoustic composition. It has a broader application, as Scruton's work shows—and it is his thesis which is my main concern. In fact, Scruton and exponents of *musique concrète* differ in two ways, concerning both the definition of the acousmatic, and its application. First, the definition. To reiterate: in describing their work as 'acousmatic music', those in the *musique concrète* tradition cite the Pythagorean definition of 'acousmatic' as 'listening without seeing'.[24] Compare this definition with Scruton's broader and subtler characterization of acousmatic listening as excluding both thought and awareness of the source or cause of the sound. On his account, such listening could occur while the cause of the sound is visible; so while both Schaeffer's and Scruton's senses of acousmatic involve detaching the sound from its circumstances of production, they should not be equated. (A third possibility is listening without knowing the cause). To clarify the contrast between Scruton and Shaeffer, recall Scruton's description of the acousmatic character of musical experience:

The person who listens to sounds, and hears them as music, is not seeking in them for information about their cause, or for clues as to what is happening... [T]he notes in music float free from their causes... What we understand, in understanding music, is not the material world, but the intentional object: the organisation that can be heard *in* the experience.   (Scruton 1997: 221)

Clearly when someone hears musical sounds they may gain information about their cause, but Scruton's claim is that in musical listening we spontaneously detach such information. This descriptive claim contrasts with the more prescriptive claim of the Schaefferians. According to Scruton, we do not have to choose to listen to musical sounds acousmatically, but do so quite naturally and spontaneously; Schaefferians—thinking in terms of 'listening without seeing'—believe that the listener has to make an effort to forget the origins of the sounds. As my later arguments show, I believe that the Schaefferians are nearer to the truth.

---

musical schema lent itself to being expressed in sound in more than one way. An example is that Bach sometimes composed without specifying the instruments: he wasn't interested in the sound of his music' (Schaeffer 1987). This questionable understanding of Bach will be addressed later.

[24] Dack (1994), for instance, writes that '[The] acousmatic situation must be extended to all those listening environments in which sounds are heard without any visual confirmation of their sources...'; Michel Chion quotes Schaeffer's definition of acousmatic sound as 'sounds one hears without seeing their originating cause', adding, '[The opposite of] visualised sound... The acousmatic truly allows sound to reveal itself in all its dimensions' (Chion 1994: 32, 71–2).

Scruton and Schaeffer differ over the application as well as the definition of the acousmatic. Schaeffer focused on our experience of non-tonal sounds or noise, which up to his time music had hardly considered, while Scruton applies his concept to what *musique concrète* composers regard as traditional music. When Scruton speaks of sounds that are detached from the circumstances of their production, he is referring to the way that tones are intentional objects of musical perception; for him, indeed, typical cases of *musique concrète* would not qualify as music. However, Schaeffer's followers at least on occasion have allowed that acousmatic experience can apply to traditional music. Bernard Parmegiani, for instance, comments that 'To analyse the sound we have to forget the source, whether it be a note on the piano, the wind, the roar of a lion'. In a piano concerto, he continues rather enigmatically, 'we never question the sound produced by the piano—it's a habitual medium, likewise the orchestra ... We forget the cause because we know it by heart'.[25] Luke Windsor (2000: 9) is more explicit, allowing that there is 'both *intentionally* acousmatic music and music that is more *coincidentally* acousmatic ...'—presumably he means *musique concrète* and traditional music respectively. Moreover, *musique concrète* compositions are indeed compositions—that is, while reduced listening involves a concern with sound itself, in which one develops a heightened attention to individual sounds, the listener must return to the whole, incorporating that new attention into the complex totality. If *musique concrète* is sound art or a precursor of sound art, it shares many of the traditional concerns of music.

The distinction between acousmatic and non-acousmatic in Scruton's sense clearly requires further elucidation. Sound experienced in terms of its cause—as the sound of some event such as the sound of a door slamming, a dog barking, or a clarinet being played—might be described as: significant, anecdotal, associative, or dramatic sound; or, conceived explicitly as a kind of experience, the purely acoustic, the practical, the literal, the documentary, the non-aesthetic. The description 'purely acoustic experience' could apply just as well to the acousmatic as to the non-acousmatic case, and so is best avoided; I will return later to the suggestion that the acousmatic is simply the aesthetic as applied to sound. 'Literal', 'practical', and 'documentary' have the right connotations.[26] Let's say I am walking in the woods and hear a creaking

---

[25] Talk given at Liquid Architecture Festival, Australian Centre for the Moving Image, Melbourne, July 12, 2003, quoted in Hamilton (2003b).

[26] The term 'literal' is used by electronic composer Trevor Wishart (1996). Schaeffer (1966: 92) writes that 'acoustic and acousmatic are not opposed like objective and subjective', suggesting that he does not want to equate the acoustic and the non-acousmatic; but the issue is not clear.

sound above me. An acousmatic response would be, 'That's a very interesting high-pitched sound, intermittent and rising in intensity'. Perhaps it could be located in Schaeffer's taxonomy of sound-objects. A non-acousmatic response, in contrast, might simply be to look up, while thinking, 'Is that a branch about to topple onto me?' Hearing is subservient to sight in information-gathering (Lippman 1999: 26–39). The acousmatic experience of sound excludes its literal qualities—as in the case of music, the listener detaches the sound from its worldly source or cause. In contrast, literal experience of sound involves a practical or technical interest. Rescuers listening for the cries of survivors buried by an earthquake treat those sounds practically and not acousmatically; a sound engineer's concern with a recording may be literal in contrast to that of a musician. Medical students are taught to listen for certain rhythmic patterns in a heartbeat, and their listening must be non-acousmatic; they are searching for information—for symptoms of a disorder. The important suggestion that non-acousmatic experience is essentially multi-modal—that it involves senses other than hearing—is discussed in the final section below.

Care must be taken to demarcate the 'cause' of the sound. In the case of *musique concrète*, audiences of course realize that the proximate source is a sound system which they can see, but this is not the cause which concerns Schaefferians. Attention to more distant causal processes can become a technical concern with how the sound is produced—and clearly, to attend to the cause of a barking sound does not require an understanding of canine physiology. I can hear an unfamiliar engine noise and not have any idea about the fault in my car that is causing it; but I do at least attend to the engine. The paradox of *musique concrète* is that the source-sounds are worldly and therefore difficult to hear acousmatically, yet when subject to electronic manipulation and treatment, experience of them becomes quintessentially acousmatic. Perhaps in contrast to the sounds of pure electronic music, they defy clear attribution precisely because we know that in their original state, their attribution is only too clear—we know that in the genre of *musique concrète* there will be a worldly source and our inability to identify it may be frustrating or perplexing.

Is it really possible to experience non-musical sounds acousmatically, as I have just assumed? The raison d'etre of *musique concrète* is that it is possible, while the impression given in Scruton's *The Aesthetics of Music* (1997) is that it is not. However, in his contribution to the present volume, Scruton (Chapter 3) takes a different view. He writes that

our language for characterizing sounds tends to describe them in terms of their normal source—dripping, croaking, creaking, barking. But reference to a source is not essential to the identification of the sound, even when it is compelled by the attempt to describe it. It is in some sense an accident if we can attribute a sound to a particular. ... (xx)

He later continues:

the kind of 'streaming' that goes on in musical hearing is [not] the same as the streaming of ordinary sound perception... since it is shaped by spatial metaphors that are the product of a musical imagination. Nevertheless, music is an extreme case of something that we witness throughout the sound world, which is the internal organization of sounds as pure events, [detached from a cause]. (xx)

Such a position opens the way to a genuinely non-musical sound art—not something close to Scruton's heart, one imagines—with elements describable by Schaeffer's taxonomy of non-musical sounds. Certainly it is the case that sound phenomena which are not music or sound art have acousmatic—one might say musical—aspects, such as the rhythm of a train engine or the melody of speech patterns. A heartbeat is a natural rhythm, birdsong is melodic; nature can be musical, even if it is not music, which has to be an intentional production.

4. Objections to the Acousmatic Thesis

The acousmatic thesis faces strong objections, and my conclusion is that it cannot be sustained—both the acousmatic and the non-acousmatic are essential aspects of musical experience. The issue is a subtle one, however, and needs careful handling. An overly sympathetic portrayal of the acousmatic thesis sees it as a twofold account which holds that experience of music must be both acousmatic and non-acousmatic. However, in Scruton's version, only the acousmatic is a genuinely musical aspect of musical experience. So the question is not whether ordinary musical listening involves attention to cause—instrumental or vocal medium—and melodic, rhythmic, and other aspects, but whether both of these are fully musical aspects of musical experience. I do not say that Scruton denies that there are these two aspects of listening. My argument is rather that he wrongly denies that the non-acousmatic aspect is genuinely musical. The most important objections to the acousmatic thesis, properly understood in this way are these below:

(1) *Timbre*. In its everyday sense, timbre is, precisely, the quality or tone color of a musical note which distinguishes different types of musical instrument, or the individual qualities of different vocalists. Timbre comprises those qualities of a musical sound which relate it most directly to its source, even if pitch,

rhythm, and harmony also do so to some extent—a high pitch is not normally produced by a tuba, for instance, though when it is, it has a special timbral quality. Experience of timbre must therefore be regarded as non-acousmatic; if it is an essential part of musical experience, as it surely must be, then the acousmatic thesis is undermined. When listening to a piano concerto or a jazz pianist, one cannot help thinking 'piano', and so it is essential to the musical experience that one attends to its causal origin.

Some elucidation of the concept of timbre is required to respond to this objection. Timbral qualities include resonance—in the case of voiced sounds, for instance, the quality imparted by the action of the resonating chambers of the throat and mouth and nasal cavities—harshness, roughness, mellowness, nasality, reverberance, shrillness, and stridency. Acoustically these qualities are 'impure'. Pure tones—those exhibiting only a fundamental frequency—can occur in music, but most musical tones are composites of partial vibrations of the vibrating body as well as vibrations of the whole mass. A typical violin tone is relatively rich in overtones while the flute, stopped diapason organ pipe, and tuning fork produce a tone of greater purity–though even these have noise elements, such as the breathy sounds of the flute. Although the development of analytic listening skills will enable one to hear overtones within a musical tone, most people listen 'holistically', and recognize only a more or less rich tone quality within the fundamental pitch.[27] The final chord of a string quartet may be heard holistically as a single 'tone', all four instruments fusing to create a complex sound; or at the analytic extreme, the individual contribution of the first violin could be further heard as a fundamental plus a large number of overtones and partials.

There are two kinds of response to the timbral objection. One is to bite the bullet and deny that experience of timbre is a central and fully musical part of musical experience; the other is to deny that timbre has to be experienced non-acousmatically. The second response involves the revisionary claim that timbral qualities may be described—and so with practice heard—acousmatically, under a description such as 'exhibiting such and such a waveform'. It did not take Schaeffer's exhaustive investigations to show that wave-form shapes—triangular, saw-tooth, sine-wave, square—offer a crude characterization of different timbres. But especially after the example of his taxonomy, it could be argued that characterizing timbre in terms of the instrument or voice that produces it is just a matter of convenience or convention; those who work in audio may struggle to characterize timbre acousmatically, but the difficulty may be a practical one. Since timbre turns out

[27] The contrast between analytic and holistic comes from Sethares (1997: 26).

to involve a complex of auditory phenomena, however, the revisionary proposal is not straightforward. Without endorsing the deflationary concept of timbre as 'the psychoacoustician's waste-basket'—a heterogeneous collection of factors, some of which arise from the fact that we often hear what we believe we hear—one should question the simple, unifying claim of spectral theorists that timbre is not essentially distinct from pitch, and is unified with it in a sound-spectrum.[28] The distinction between tone and harmonic spectrum, for spectralists, depends on whether the fundamental tone or the harmonic relationships of the spectrum are given more attention. This unifying account seems to focus on the 'steady state' portion of the sound, and downplays the temporal aspects of each instrument's characteristic frequency envelope—attack, decay, sustain, release, and transients. The attack—the initial production of the note, such as the hammer hitting the piano strings or the first blat of the trumpeter's lips—is highly characteristic, and removing it from a piano or trumpet recording makes it much harder to identify the instrument. Noise content is also important, as Thomson stresses:

> The friction of the bow as it is set into motion across the string, the eddies of air pressure within a horn's mouthpiece, or the hammer's impact on a piano string [are important]...After articulation, however, it is the presence or absence of overtones and their relative intensities that determine...timbre. (Thomson 2004)

Attack and noise are the least acousmatic elements of timbre.

The complexity of timbre makes it implausible that one could conceptualize it in terms of wave-form pattern, instead of in causal terms as 'violin' or 'saxophone'. This negative conclusion is reinforced by the difficulty of synthesizing credible instrumental tones—attempts to simulate the wave-form of a violin sound with an oscillator or other tone-generator, rather than by recourse to sampling, remain unconvincing. The response constitutes too radical a revision of traditional music practice. It cannot be just a matter of convenience or convention that timbre is characterized in terms of the instrument or voice that produces it. Beethoven in his string quartets asks for a cello, not a cello-like sound—the non-acousmatic element is essential.[29]

---

[28] The spectralist view is expressed by Smalley when he writes: 'The idea that "notes", the bearers of pitch information, are clothed in timbral hues is not eliminated, [but] located in the wider perspective of spectral types' (Smalley 1986: 65). The crucial step from harmonic to inharmonic spectra or noise required electronic technology to become viable. Spectral composition is discussed in Harvey (1999: ch. 3) and in Hamilton (2003a).

[29] This is not quite the question whether a particular instrumentation is essential to a work—on which see for instance Davies (2001: 47–71), which argues that a work's instrumentation is essential only from the time when composers could specify such things and expect performers to take notice, namely the 18th century.

(He was not in a position to ask for a cello-like sound not produced by a cello, but the point remains.) The only music which could properly be experienced acousmatically in this sense is music in which instrumental specification is arbitrary—such as Bach's, according to traditional scholarship, or, more recently, Anthony Braxton's—or electronic music which avoids traditional instruments.[30] Parmegiani's call, quoted earlier, to forget 'piano' when hearing a piano concerto is unwarranted as well as unrealistic. The revisionary proposal is also undermined by a different response to the timbral objection, which I will consider later in this section.

Scruton's response to the timbral objection is to bite the bullet and deny that experience of timbre is a central and fully musical part of musical experience. For him, timbre is not one of the fundamental elements—pitch, rhythm, melody, and harmony—by which sound is organized into tone, and he grounds it in the character of sounds rather than in their organization. Scruton's neglect of timbre perhaps reflects a realization of its threat to the acousmatic thesis, but it also expresses his opposition to key aspects of modernism. For while timbre was secondary to pitch in the pre-modernist history of Western art music, modernism, in the form of Schoenbergian *klangfarbenmelodie* and more importantly Debussy's mature work, elevated timbre to structural status. In Debussy, motivic particles replace themes; harmony ceases to be an agent of musical motion and instead becomes static, atmospheric, and coloristic; and texture, color, and dynamic nuances assume unprecedented importance. It is often argued that there is a concern with sound for its own sake (Morgan 1991: 42–6). Edgard Varèse, greatly influenced by Debussy, had a more abrasive conception of timbre; for him, it has been said, sound was 'a physical phenomenon...[T]here was always [a] very palpable sense of sound as vibration' (Clark 2005: 40). Varèse commented on how 'color or timbre would be completely changed from being incidental, anecdotal, sensual or picturesque; it would become an agent of delineation, like the different color on a map separating different areas, and an integral part of form' (Cox and Warner 2004: 18). A more recent illustration of the centrality of timbre is found in spectral composition, which like *musique concrète* grew out of the French musical tradition, though many of its foremost practitioners have come from elsewhere; composers such as Grisey, Radulescu, Harvey, and Saariaho, building on the intuitive deployment of tone-color by Debussy,

---

[30] The received view of Bach's 'arbitrary' instrumentation needs qualification. The fact that it is a solo violin for which he writes harmonic music is essential to the musical experience; what could be played effortlessly on the organ becomes something almost miraculous on a four-stringed instrument. Another example is his profound sensitivity to the different ranges of the human voice, reflected in his settings of liturgical texts for specific vocal registers.

Varèse, Messiaen, and Boulez, have exploited the harmonic spectrum in a systematic microtonality. (Hence 'spectralism'.)

Timbre also has an important structural role in some non-Western music, such as the Japanese *shakuhachi* tradition, where as an organizing principle it may be at least as important as pitch or rhythm; a piece may be regarded as a series of timbres as much as pitches (Lee 1988). However, the recognition that timbre can have structural status both in modernist and non-Western music in fact helps the acousmatic thesis, since it is precisely the character of sounds which relates them to their causal origin, while organization divorces sounds from their cause. So in such music, timbre is experienced acousmatically. A dilemma thus opens up for proponents of the timbral objection: either timbre has a structural role, and is experienced acousmatically, or else it is musically less significant, and experienced non-acousmatically. The timbral objection is therefore not decisive in refuting the acousmatic thesis.

(2) *Space*. It turns out that the same dilemma confronts the objection that acousmatic experience cannot involve awareness of the spatial origin or movement of sounds, which clearly concerns their cause or source. Acousmatic experience, the objection continues, is not sufficient for the appreciation of those kinds of music which aim to achieve spatial effects through placement of groups of performers or sound-producers—where it is important that one attends to the direction of the sounds.

Despite the examples of Baroque antiphonal music and 19th century compositions with off-stage musicians, such a purpose was not prominent before the 20th century. But it is central to contemporary compositions beginning with Stockhausen's *Gruppen* and *Carré*, though prefigured in some ways by the work of Charles Ives. Almost from the start of his career, Stockhausen wished to undermine traditional concert hall listening. Describing *Gruppen* rather grandiosely as the first example of 'spatial music', he demands specially designed halls with moveable seating to allow his music's 'theatrical polyphony' to be realized:

> The function of space has been neutralized [in the Western tradition] ... [M]ost of the audience can't even stand, let alone move during a concert, so our perspective on musical space is utterly frozen ... If I have a sound of constant spectrum, and the sound moves in a curve, then the movement gives the sound a particular character compared to another sound which moves just in a straight line. Whether a sound moves clockwise or counter-clockwise, is at the left back or at the front ... [are] configurations in space which are as meaningful as intervals in melody or harmony. (Stockhausen 1989: 101–3)

Sonic architecture in Stockhausen's sense did not figure in the traditional craft of the musician, nor was it part of the necessary equipment of the listener.

As Trevor Wishart (1996: 136) comments, 'the control and composition of landscape [the experienced source of the perceived sounds] open up large new areas of artistic exploration and expression'.[31] However, it does not seem entirely correct to say, as Wishart does, that artistic control of landscape or soundstage depends on the advent of recording. As the example of Baroque antiphony shows, sound recording fosters a concern with landscape, but landscape is not essential for it; exploitation of stationary, dispersed sound-sources, and indeed moving sound-sources, is possible without it. Thus a trumpeter, strapped in a harness attached to a high-wire and playing their instrument while propelled across the auditorium—a trick that admittedly will require some practice—gives a real impression of a moving musical sound, including Doppler effect. Non-electronic compositions such as Xenakis's *Metastasis* and Stockhausen's *Carré* have created an acoustic illusion of this kind of movement, but electronic composition achieves an imaginative leap in such possibilities. The plasticity of the sounds—their malleability and manipulability as objects—and the vividness of sound projection can generate a brilliant impression of their propulsion across the soundstage.[32]

However, it is not clear that the experience of spatial effects, integral to such music, is non-acousmatic—for the same reason as in the case of timbre. Since organization divorces sounds from their cause, in music where spatiality has structural status, spatial aspects are experienced acousmatically. Sonic landscapes are strictly illusory artistic representations or mediations of the anecdotal or literal, and so one does not attend literally to the causal origin of the sounds.[33]

---

[31] Wishart (1986) breaks down perception of landscape into three components: the nature of the perceived acoustic space, the disposition of sound-objects within that space, and the recognition of individual sound-objects. In a recording or electro-acoustic composition whose landscape is a forest with sounds of birds and animals, he distinguishes three ways of disposing sound-objects: 'unreal objects/real space' (animals and birds which cannot exist in close proximity, or animal and bird sounds replaced by arbitrary sonic objects), 'real objects/unreal space' (original animal and bird sounds assigned different amplitudes, reverb or filtering to create a different kind of imaginary landscape), and a surrealist 'real sounds/real space' in which the relationship between sound-objects is impossible (an extreme version of 'unreal objects/real space'—though it seems hard to distinguish from the former).

[32] There are many examples of the control of 'landscape' or soundstage in contemporary electronic composition. Jonathan Harvey's 'Mortuos Plango, Vivos Voco' is a modern classic of the genre, and special mastery of landscape is shown for instance in Rolf Wallin's recent 'Phonotype 2'.

[33] In fact, Stockhausen denies that landscapes generated by electronic music are illusory: 'Our conception of truth of perception is entirely built on the visual... [hence] most people listening to [electronic] music... when they hear the sounds in a given hall are moving very far away, and coming very close, they say well, that's an illusion... We now have the means technically to make the sound appear as if it were far away: "as if", they say. A sound that is coming from far away is broken up and reflected by the leaves of the trees, by the walls and other surfaces... A sound that is very close to my ear reaches my ear directly, without reflections, and [this] can also be produced artificially... [Now] when they hear the layers revealed, one behind the other, in this new music, most listeners cannot even perceive it because they say, well, the walls have not moved, so it is an illusion. [We need to

As with timbre, however, this still leaves the case of music where spatiality lacks a structural role. While traditional concert music involves the construction of a soundstage—for instance through orchestral layout—this process is not a means of artistic expression, since the concern is simply with clarity and balance rather than the exploitation of spatial effects. There is a parallel with the traditional rectangular cinema screen format; since this is the norm, only divergences from it can be artistically expressive. Nonetheless, one can appreciate how both the improvising and non-improvising musician respond to the room's acoustic, adjusting their performance to its idiosyncrasies; this is an aspect of one's enjoyment of the music which is non-acousmatic.

(3) *Virtuosity*. Acousmatic experience cannot involve awareness of virtuosity in performance, so it will not allow for appreciation of music where this is a significant element in the listener's appreciation. It is part of one's experience of Liszt's *Transcendental Studies*, Louis Armstrong's 'Swing That Music', or Ferneyhough's 'Unity Capsule', that these are technically extraordinarily difficult; a recording of Liszt's pieces where the right-hand part was overdubbed using two hands would lose the elements of devilry, risk, excitement, and relief. This point applies to other expressive qualities, too: the sense of strain generated by the first violin part in the 'Cavatina' of Beethoven's opus 130 string quartet—which calls for very high positions on the lower strings of the instrument which are very taxing for the performer—is an intrinsic part of that quartet's expressive power. These qualities are apparent audibly, but they are also apparent visually, and thus give rise to the final objection below.

(4) *Experience of music is not purely auditory*. We feel as well as hear sounds; indeed, some music seems to emphasize this fact. As noted earlier, in the music of Varèse, a sense of sound as vibration is integral. The fact that sounds are felt allows exceptional hearing-impaired musicians and dancers, such as Evelyn Glennie, to make a career; rhythm does not have to be sonic when expressed in bodily movement.[34] As we saw earlier, Schaeffer's concept of the acousmatic was built on the assumption that where sight is involved, it is difficult and perhaps impossible to experience sounds while abstracting them from their causal origin. When we witness a musical performance, experiencing musical sounds as humanly produced seems inescapable. We see, as a direct causal process, how the music is energized by the actions of performers. The visual

---

believe] in what we hear as absolutely as we formerly believed in what we see or saw...What makes it so difficult for new music to be really appreciated is this mental block' (Stockhausen 1989: 107–8).

[34] The vibratory sense is closely connected with hearing, and is discussed in Lippman (1977, 1999). The claim concerning rhythm is considered in Hamilton (2007b: ch. 5).

aspect of performance creates tension, as when we see the percussionist raise the beater to strike the drum, or a pianist perform a daring leap. The gyrations of the conductor and pianist are vital to the audience's comprehension, and an accent accompanied by an outflung arm seems to become more intense. Many of these effects arise through music's primitive connection with bodily gesture, and especially dance.[35] But electronic music creates its own kind of tension since listeners cannot prepare themselves mentally for the sounds that will occur. What is seen—or not seen—affects what is heard. (Clearly I would not argue that non-acousmatic experience always involves senses other than hearing; this issue is taken up in the final section.)

All of the preceding objections to the acousmatic thesis arise from the fundamental fact that music is an art of performance. We do not attend musical events simply to experience an auditory realism or perfection unattainable through recordings; rather, we want to see as well as hear the creation of musical sounds.[36] So, the acousmatic thesis, in the strong form in which Scruton presents it, seems obviously false. This reaction is premature, however. There is a further defense of the thesis, one which involves a broadening of the concept of the acousmatic. Consider again the timbral objection (1). According to this further defense, in listening to a piano concerto one abstracts from the particular cause but not the general one—one experiences the sounds as those of a piano, but not necessarily of *that* piano, the particular instrument causing the sound. To experience the sounds as those of a particular instrument is to adopt the attitude of a piano-tuner, piano salesperson, piano-maker, or pianist in professional mode, looking for the best instrument to use—an aesthetic concern, certainly, but only indirectly an aesthetic concern with the improvisation or work being performed. (In the case of the salesperson, aesthetic judgment is totally at the service of getting and giving information about the quality and market value of the piano.) Acousmatic experience can be disrupted by such experience of the particular cause—thus we say that an out-of-tune piano spoils one's pleasure in the music. It is true that the fundamental aesthetic concern with a particular piano is that, at least potentially, it allows the performer to make good music. But the artistic order of importance in the case of composition is: first, the work; then, the performance or interpretation; and finally, the recording (if any).

---

[35] A claim defended, for instance, by Lippman (1977: ch. 7) and Nettl (2001); it is discussed in Hamilton (2007b).

[36] It is notable that Walton (1988) omits the dimension of performance from the non-abstract features of music, when surely it is of the first importance.

The result of this line of argument is a *qualified acousmatic thesis*: to hear sounds as music is to abstract from their particular cause, but not necessarily from their general cause. As it stands, however, this qualification is not sufficient. The experience of a general cause—'piano' but not 'that piano'—can occur with improvised music as well as composition; but in instrumental and especially vocal improvised music, the particular and not the general cause seems essential to musical experience. When I am listening to the singing of Billie Holiday, Mose Allison, or Bob Dylan, it is part of my musical experience and enjoyment that I do not abstract from the particular cause, and do not abstract from its production by a particular individual. I do not for instance experience the sounds as just having some generic cause—for instance, 'African-American jazz singer'. It may well be that vocal sound, with its intimate connection with the performer's body, emphasizes the non-acousmatic.[37]

There is a further development of the acousmatic thesis, which, though elusive, offers a response even to this seemingly decisive objection. It is based on the claim of a necessary metaphorical perception of music—which in one form is central to Scruton's treatment—and argues that the perception of 'piano' is metaphorical rather than literal, and that the content of perception is culturally mediated. This consideration applies to experience of general and particular causes alike. A vivid example is the experience of piano sound. Keyboard music of the Classical and Romantic eras, from Haydn up to Bartók, in fact, which has decisively shaped our perception of the piano, is so concerned with the projection of a legato sound that it makes us forget that true legato is impossible on the instrument. In the case of strings, brass, and woodwind, the envelope of sound is extinguished as the player moves from one note to the next, without overlap, and, if they wish, without any gap. This is not possible with the piano. However, listeners ignore the decay of the piano's sound, and accept the aspiration to a perfect legato as real. The gestures of the pianist help to sustain this illusion—and so the visual experience of an illusory cause supports rather than opposes the acousmatic thesis. Charles Rosen comments that

> we hear the sounds in a Beethoven piano sonata as if they were sustained by string instruments or voices... More than any other composer before him, Beethoven understood the pathos of the gap between idea and realization, and the sense of strain put on the listener's imagination is essential here. (Rosen 1999: 2–3)

A piano of Beethoven's time is ideal because of its greater inadequacy for conveying such an effect, he continues, but adds wryly that the modern piano

---

[37] Lippman argues that 'the solo voice has a pronounced externality [non-acousmatic nature]... [A]n individual singer remains an object that insists on an external status' (Lippman 1977: 70).

is sufficiently inadequate to convey Beethoven's intentions. The sound of the piano is perceived metaphorically, as akin to a legato string sound, even though on reflection, listeners recognize that the instrument cannot really produce this sound. The phenomenon is one of the mysteries of Western art music.

There is another kind of metaphorical perception which does not, strictly speaking, involve illusion—the experience of the residue of historical practice, for instance in declamatory French horns in a Bruckner symphony. Here one immediately perceives a horn-call, with its connotations of hunting and the chase; the sounds are heard as organized into a distinctive musical phrase, with conventional or cultural connotations. Of course this is not literally a horn-call, but a cultural construct derived in part from historical practices such as hunting. Even sounds which appear to call particular attention to their causes, such as 'dirty' sounds—the bottle-slide technique of the blues guitarist, or the growling or squeaking noises produced by a free jazz saxophonist—are objects of metaphorical perception. Dirty sounds tend to be stylized imitations of the human voice by an instrumentalist. It is perhaps no coincidence that Western art music, with its profound structural concerns, has a preference for pure rather than dirty sounds—hence for instance the alien nature of John Cage's prepared piano sounds, which he intended as subversive.[38] But still, to say that dirty sounds are stylized is to regard them as possessing a cultural overlay—to regard them as objects of a metaphorical perception. Although the preceding considerations would be cited by proponents of the 'psychoacoustician's waste-basket' conception of timbre mentioned earlier, which I would not endorse, we do often hear what we believe we hear. (Michel Chion's discussion of cinema sound, and the McGurk effect, make the same point, on which more below.)

5. The Twofold Thesis

The preceding responses, I believe, are not sufficient to preserve the acousmatic thesis. There is a literal as well as metaphorical dimension to musical experience; but since the claim of necessary metaphorical perception does not strictly imply that all aspects of musical perception are metaphorical, the latter thesis is both weaker and more plausible than the acousmatic thesis. The acousmatic thesis is too prescriptive about what musical experience involves. Part of the pleasure in listening to music is a sensuous pleasure in sounds, which may not involve

---

[38] As Jonathan Harvey put it, 'A smooth unchanging stream of neutral timbre invites attention mainly onto metaphysical events' (private communication—and see below).

acousmatic experience. There is also a further issue. Attempts to broaden the acousmatic—such as the qualified acousmatic thesis described earlier, which admitted experience of causes in a general sense—may succeed simply in equating 'acousmatic' with 'aesthetic (as relating to sound)'. According to this view, musical experience is not experience of sound divorced from its cause; rather, it is experience which involves an aesthetic attitude towards its cause. A Kantian conception of the aesthetic holds that aesthetic experience of sounds divorces them from their original context, and does not treat them as providing information. If the acousmatic is interpreted as 'not involving an interest in information about the cause of the sound', it then becomes the Kantian aesthetic as applied to sound. This is a definition towards which Scruton himself sometimes inclines.[39] Although it must be the case that genuinely musical experience is essentially aesthetic, the upshot is that the acousmatic ceases to be distinctive.

In place of the acousmatic thesis, I propose a *twofold thesis*, which states that listening to music involves both non-acousmatic and acousmatic experience, and that both are genuinely musical aspects. (In the case of singing, which involves a text, the experience becomes threefold; one can listen non-acousmatically to the voice, attend to its musicality acousmatically, or focus on the meaning of the words.) This twofold thesis is implicit in the work of various writers. For instance, Thomson comments that tone differs from noise mainly in that it possesses features that enable it to be regarded as autonomous:

> Noises are most readily identified, not by their character but by their sources; e.g., the noise of the dripping faucet, the grating chalk, or the squeaking gate. Although tones too are commonly linked with their sources (violin tone, flute tone, etc.), they more readily achieve autonomy because they possess controlled pitch, loudness, timbre, and duration, attributes that make them amenable to musical organization. (Thomson 2004)

And composer Jonathan Harvey writes: 'One is constantly alternating as a listener between delight in the sound and delight in the structure, depending on the composer's emphasis (and the player's)'.[40] Even Pierre Schaeffer's sympathizers have recognized that musical experience is twofold in this sense. In his investigation of *musique concrète*, Luke Windsor (2000: 9) holds that 'for

---

[39] For instance in the passage quoted earlier: 'The person who listens to sounds, and hears them as music, is not seeking in them for information about their cause, or for clues as to what is happening' (Scruton 1997: 221).

[40] Harvey adds: 'A lot depends on changes in articulation. If Beethoven was the first for our ears to emphasise violent changes of playing articulation, he is also the most obvious start in history to "listening to sound". A smooth unchanging stream of neutral timbre invites attention mainly onto metaphysical events' (email communication with the author).

the listener at least, attempts to break through the acousmatic "screen" in order to ascribe causation to sounds are an important facet of musical interpretation'.

The analogy with Richard Wollheim's twofold thesis of 'seeing-in', concerning the experience of pictorial representation, is deliberate. Wollheim's claim is that one experiences a picture non-representationally and atomistically, as a set of marks on a surface, and also representationally. For Wollheim, 'seeing-in permits unlimited simultaneous attention to what is seen and to the features of the medium... [I]f I look at a representation as a representation, then it is not just permitted to, but required of, me that I attend simultaneously to [pictured] object and medium... though of course [my attention] need not be equally distributed between them...' (1980: 213). The analogy with musical experience is that just as looking at a painting involves experiencing or being involved in both the represented scene (the Nativity or the peasant's boots) and the means of representation (paint-marks on canvas), so listening to a piece of music involves experiencing the sound as part of a musical world of tones, and as having physical properties and origin. This twofold quality reflects the contrast between atomistic and holistic experience—the acousmatic is holistic experience of musical structure, while the non-acousmatic is atomistic experience of individual, merely causally related and meaningless sounds.

Now, as noted earlier, the question is not whether ordinary listening involves attention to both cause or medium and tonal aspects, but rather, whether each is a fully musical aspect of musical experience. My objection is not that Scruton rejects twofold-ness, but that he wrongly denies the genuinely musical status of the non-acousmatic aspect. He does indeed seem to hold with Wollheim that there is a single act of attention. Thus, while Wollheim argues that I must be able to see the cornfield in the picture in the same act of attention that reveals to me that it was produced by means of a pallet knife working on chrome yellow paste, Scruton argues that I must be able to hear the phrase that opens the second movement of Brahms's Symphony No. 4 as a melodic unity, at the same time as hearing that it is sounded on the horns. According to this concept of 'double intentionality', acousmatic experience is available in one and the same act of attention that embraces the real-life causality of the musical medium; we focus on something real while attending to something that is imagined in and through it.[41]

Perhaps Scruton is right to suggest that one should not or cannot pick out aspects like spatial properties and timbre, and treat them as non-acousmatic, as if they could be the object of a distinct act of attention. The issue is not clear.

---

[41] Scruton argued this in discussion.

But in any case, double intentionality does not, as he seems to assume, offer a kind of proof of the acousmatic thesis, for the reason that, as I have been arguing, 'real-life causality' is a genuinely musical part of musical experience. For 'genuinely musical' here, one could substitute 'genuinely aesthetic'. The genuinely musical is not entirely imagined, entirely the product of metaphorical perception, or essentially acousmatic. Nonetheless, it seems to be a consequence of the nature of sound that mediation through the concept of causality has a particular significance in musical experience that it may not have in arts such as painting and literature.

It is undoubtedly the case, for instance, that a more developed musical understanding tends towards the acousmatic. We say that the playing of a novice musician is beginning to make musical sense, that it is becoming less mechanical—we can experience it more acousmatically, though the mechanics of sound-production may later be exploited intentionally, as effects. (For instance, contemporary avant-gardist Helmut Lachenmann's extended instrumental techniques draw attention to the means of their production—one of his primary reasons for exploiting them).[42] Likewise, a novice's appreciation of music is less acousmatic; a more developed understanding becomes more predominantly acousmatic. The novice in a particular genre—Western listeners confronted with Tibetan throat-singing, for instance—will want to know how the sounds are produced. If someone cannot recognize what instrument is being played, to that extent their musical experience is impoverished, even though they may be able to give a quite detailed description of the kind of sound—reedy, nasal, or whatever.

We saw earlier that there is an interesting tension in Scruton's account between his advocacy of the acousmatic thesis, and his assumption of what may be described as a humanistic conception of music, which stresses its origins in dance, ritual, and gesture, and its connection with human life and activity. The autonomy of musical sound from its causes, which tends to separate music from the world, is one of several reasons why music is held to be the most abstract of the arts. The acousmatic is not the only dimension of musical abstraction, however. An abstract conception is implicit in the ancient Pythagorean concept of music as number, and received an impetus from the development of the work-concept in the 18th and 19th centuries, and the contemporaneous decline of the language-model

---

[42] Lachenmann, through his concept of *musique concrète instrumentale*, explores new possibilities of sound production using traditional instruments. The singing instrumental tone, which he regards as 'domesticated by tradition', is replaced by 'the detritus of sonic phenomena', with a preponderance of toneless sounds, mostly breathing, from wind instruments, and brutal grinding and scraping of the strings (Heathcote 2004).

and rise of absolute music. 'Abstraction' proves an elusive target, however, in part because the opposition between abstract and humanly concrete is dialectical in Adorno's sense—the opposites turn out to be overlapping or inter-penetrating, not diametrical. Thus what at first sight appears abstract in fact has concrete, sensuous, or meaningful elements. So although Scruton's humanistic conception appears to be in tension with his advocacy of the acousmatic thesis, the concept of musical form that arises from the acousmatic need not be abstract.[43]

Totally synthesized electronic music, such as Stockhausen's early studies constructed from sine-tones, amounts to the limiting case of acousmatic abstraction considered in this chapter. *Musique concrète* draws music and life together by using everyday sounds as its material, and even electro-acoustic composers apparently committed to abstraction have adopted a humanistic standpoint. Jonathan Harvey, for instance, notes that sounds in electronic music often have only vestigial traces of human instrumental performance—no one can be envisaged blowing, hitting, or scraping anything—but maintains that this process has necessary limits. Citing music's relation to rhythms of heartbeat and breathing, and to our sense of gesticulating, walking, running, and dancing, he declares it 'onomatopoeic through and through' (Harvey 2008; 1999: 57, 62).[44] It follows that acousmatic listening, which abstracts from causes, constitutes an idealized approach even to electro-acoustic music. Music's autonomy from its causes is an incomplete one.

6. Music and the Metaphysics of Sound

There is much more to be said on the question of the twofold thesis. However, I will conclude with remarks on some implications of the preceding account of acousmatic experience for the metaphysics of sound. In his contribution to the present volume, Scruton develops the anti-physicalist position that sounds are secondary objects and pure events, and argues that this metaphysical status underlies the art of music. A secondary object is one whose properties are entirely comprised of ways in which it appears—yet which is not merely a subjective impression, but part of the objective world. A

---

[43] Adorno's (1997) 'negative dialectics' are classically presented in his *Aesthetic Theory*, and discussed in Hamilton (2007b).

[44] Dennis Smalley writes that 'we detect the humanity behind [abstract musical structures] by deducing gestural activity'—presumably, gestures associated with the production of sounds: 'Music is always related in some way to human experience' (Smalley 1986: 64).

pure event is one which does not happen to anything, and which cannot be reduced to changes undergone by re-identifiable particulars—in contrast physicalism concerning sound, reduces sound to the vibration of its source. Tastes and smells are other examples of secondary objects and pure events, Scruton argues. He claims that the essentially metaphorical nature of musical experience obliges us to treat sounds in this way. His metaphysical position should be contrasted not only with the physicalist identification of sound and mere vibration, and with claims of reification encouraged by recording technology.[45] It should be contrasted also with the more radical *non-spatiality thesis* that sounds have spatial properties only contingently, and need have none at all—that sounds are objects which are not part of the material world.

This anti-materialist thesis appears to find a basis in Strawson's metaphysical *tour de force* of a purely auditory world, presented in the second chapter of *Individuals* (1959). Strawson's conclusion is that purely auditory experience alone cannot furnish objective concepts of a spatial world. One can experience sounds as located, only if one has already acquired objectivity-concepts from the other senses (Strawson 1959: 65). Matthew Nudds (Chapter 4; 2001) interprets Strawson's perhaps injudicious claim that 'Sounds...have no intrinsic spatial characteristics' (Strawson 1959: 65) as the thesis that sounds could lack spatial properties altogether—the *non-spatiality thesis*. As Nudds says, we are not auditorily aware of empty places—there is no difference between not experiencing a sound at some place, and experiencing no sound there. As further evidence for the non-spatiality thesis, he claims that it is possible to attend to the sound of the oboe in an orchestral context without being able to distinguish its location from the location of other instruments; and one can experience ringing in one's ears without hearing it as having any spatial properties. Thus, 'We can hear a sound without hearing it to have any spatial properties at all...Not only can we imagine a world of sounds which is a no-space world but...we can imagine the actual world of sounds as a no-space world.' He concludes that while there appears to be a unity to the objects of sight and touch, 'sounds appear not to be part of [the material world]' (Nudds 2001: 213–15).

The non-spatiality thesis involves a radical affirmation of the acousmatic. Indeed, Edward Lippman links the fact that humans are not good at locating

---

[45] For instance, Robert Worby's comments that 'Recording makes sound tangible, concrete, almost tactile. It becomes data on a storage medium...[C]omposers have [used it] to make sound tactile and malleable, like paint or clay...What photography and sound recording enabled was the freezing of the moment...[turning] something that was fleeting, evanescent and transitory...into a thing, an artefact' (Worby 2004).

aurally the sources or producers of sounds with the acousmatic experience of music:

> It is... the relatively poor capacity of the ear to locate sources of sound, even if it is aided by head movements and vibration, that is partly responsible for the peculiar prominence of immanent auditory objects, which have no location in environmental space and no really conformable counterpart in empirical experience or physical science. (Lippman 1977: 69)

It might be argued that source-location is now an almost vestigial function of hearing—in contrast, our ancestors had moveable ears—but it is also interesting to note how reliant on hearing people must have been in the not so distant eras before electric lighting. And I still wish to maintain my argument against the acousmatic thesis, that musical sounds are essentially part of the human and material worlds. Therefore I wish to resist the non-spatiality thesis, which I believe involves a misrepresentation of Strawson's position. Strawson was right to argue that hearing alone is insufficient to furnish objectivity-concepts. But this does not mean that—in a world of subjects equipped with the normal range of human sensory capacities—sounds have no intrinsic location. 'On the basis of hearing alone' is ambiguous between 'possessing other senses but not using them on this occasion' and 'never having exercised other senses (and so not having developed objectivity-concepts)'. When Strawson writes that sounds have no intrinsic spatial characteristics, he is referring to the latter situation, not the former; he means that on their own, without the assistance of the objects of sight and touch, sounds cannot furnish spatial concepts. He would, I think, allow that for a subject who does possess the normal range of human sensory capacities, hearing alone can yield knowledge of the location—or at least the apparent location—of sounds.

If the non-spatiality thesis is rejected, what does the attribution of spatial location to sounds involve? The spatial location of a sound is where it appears to come from. This is an objective question, answers to which are given by descriptions of sonic landscape, such as those of a recording's soundstage—the way the instruments in a group appear to be spread across the area in front of and between the loudspeakers. The saxophone appears—objectively—to come from between the two loudspeakers, for instance. So the apparent location of the sound is distinguished from the location of the sound producer, although normally, the sound appears to come from where the sound producer is, and in the case of the stereo soundstage, the dislocation is not so marked—there is a sound producer in the close vicinity. Although we are not always good at working out the causal origins of sound with our eyes closed, therefore, sounds do generally appear to come from a location (Lippman 1999: 31). We often turn—without thought—in the direction of a sound, as when someone calls our name; blind people do not seem to be in a constant state of confusion about where sounds are coming from. Discrimination improves with training; one can, for instance, learn the difference between a recessed recording, where the orchestra appears to be at the opposite end of the hall to the listener, and a well-projected one where the listener seems to be in the front row.

Directional cues concerning the apparent location of the sound, or the location of the sound producer, depend mostly on the fact that our hearing is stereophonic. Sounds to the right of the subject will be louder in the right ear than the left, and will arrive there fractionally earlier, though the brain equalizes them and we are not conscious of the time-difference. But how easy it is to locate sounds on the basis of hearing alone depends on their pitch, timbre, and volume. Sounds with a full frequency range lose the bass frequency when coming from a distance. Higher frequency sounds such as birdsong give more information about spatial origin—that is, they more clearly appear to come from a particular direction—than low frequency ones; thus, sub-woofers in a speaker system do not have a critical placement in the room. Researchers at Leeds University have recently developed ambulance and police sirens which optimize directional cues, allowing other drivers and pedestrians to react more quickly to emergency vehicles, thus reducing delay and traffic accidents. The Leeds researchers concluded that traditional pure-toned sirens give very poor directional cues, and that for a sound to be localizable it should contain as much of the frequency range as possible—hence pitched tones should be separated by bursts of broadband noise (De Lorenzo and Eilers 1991; Withington 1998). (I referred here to sound arriving at the ears from the source; this assumes the sound waves model, according to which unheard sounds travel through the upper atmosphere, under water, and so on. According to this view, not all properties of sounds are 'ways in which they appear', as Scruton puts it. When sounds travel because they echo through the hall or because, like fire engines, we hear them first from a distance and then near, the situation is closer to Scruton's model.)

The existence of sounds of indeterminate location is debatable, except in a case when one is engulfed by a sound—such as the applause of the audience of which one is a member—and therefore the sound location cannot possibly be considered indeterminate. Low, rumbling sounds can appear to come from all around—but that again is a case of immersion, like being in the swimming pool as opposed to seeing the water from the poolside.[46] Solo instrumental

---

[46] 'If the speaker system [in an electro-acoustic concert] is cunningly devised, the sounds...will seem to come from nowhere and float invisibly around the hall like immaterial forms' (Harvey 1999: 57).

sounds tend to draw attention to their source; instruments in an ensemble do so less pronouncedly. In the orchestra, the French horn sounds most ethereal, as if coming from nowhere; a cello's location, in contrast, is easier to discern. But a listener can focus on the horn's general location, even though it is surrounded by other instruments whose relative locations are hard to distinguish aurally. It is not clear what one should conclude from Prospero's creation, in *The Tempest*, of sounds that beguile or confuse the shipwrecked crew. The fact that there are no physical sound-producers—Prospero produces them by magic—does not mean that the sounds appear to come from nowhere (unless, again, one is immersed in them). Illusions are mistakes about origins of sounds, and do not support the view that sound is not intrinsically located. A ringing in one's ears after a loud rock concert can be regarded as an auditory after-image, comparable to spots before the eyes after looking at bright lights. Furthermore, one cannot conclude that since some sounds appear to come from nowhere, then any sound can be imagined as lacking spatial properties—just as one cannot conclude from the fact of Thalamic Syndrome, in which pain is experienced as diffuse and lacking a location, that pains are not in most cases intrinsically located. Both are puzzling phenomena. In the case of sound, pitch and timbre are crucial here; as mentioned above, a cello sound has rather prominent spatial properties. Not only that, but certain instrumental sounds have either projective or recessive qualities—the tone of the oboe is penetrating, while that of the French horn is recessive, with a spreading diffuseness. (Compare how the color red seems to advance towards the observer and blue seems to withdraw (Lippman 1977: 54, 65).)[47]

My conclusion, then, is that sounds do have intrinsic spatial location. However, while rejecting his non-spatiality thesis, I am sympathetic to Nudds's associated claim that experience of the production of sounds is essentially bi-modal, a claim which has an immediate bearing on the nature of acousmatic experience. He argues that

When we see a dog bark and hear the sound it makes we don't just hear a sound as coming from the same place we see the dog barking; we perceive the dog to be producing the sound that we hear... We never simply *hear* something as producing a sound because we can't hear the sources of sounds apart from hearing the sounds that they make. (Nudds 2001: 200, 218)

---

[47] Lippman also writes: '[T]he arrangement and interplay of instruments has played an important role even in the more autonomous music of the West, in respect of both the varied character of the sounds and their contrasted physical locations... [In] the Baroque concerto, the opera buffa ensemble, or the early Neoclassicism of Stravinsky... [they] are heard as distinct physical entities with distinct locations relative to one another and to the listener, [while] in Impressionism... the instruments as multisensory objects play very little part in the conception and apprehension of the music' (Lippman 1977: 60–1).

Thus for Nudds, deaf people experience what they see in a different way to sighted people—they no longer experience the production of sounds. These claims are persuasive, insofar as they express the holistic or cross-modal nature of sensory perception—that the input of different perceptual channels cannot be absolutely separated, and so the onset of deafness, for instance, affects the content of vision. The claims are reinforced by observations on cinema sound made by Michel Chion concerning how strong visual cues override aural ones.[48] Chion comments that filmmakers in the early days of sound worried that audiences would be confused about the location of screen sounds—but it turned out that footsteps are heard as coming from the location of the actor who is walking or running, and so on, rather than as coming from the cinema sound system. We are not at all confused, in any doubt, or hesitant—we experience the sounds as coming from the onscreen event. If this experience is an auditory illusion, it is just the 'illusion' created by stereo sound reproduction, so 'image' may be a better description. Chion also discusses 'synchresis'—a combination of 'synchronism' and 'synthesis', which describes the use of unlikely sound effects for an actor's footsteps, for instance. When a situation sets up precise expectations of sound, he adds, 'synchresis is unstoppable ... In *Mon Oncle* Tati drew on all kinds of noises for human footsteps, including ping-pong balls and glass objects' (Chion 1994: 63–4).[49]

Clearly, concerning the location of sounds, vision dominates hearing. Pierre Schaeffer would agree with Nudds that non-acousmatic experience is bi-modal, while acousmatic is uni-modal ('listening without seeing'). He also downplays the possibility of purely auditory non-acousmatic experience. But it cannot be the case that we rely almost exclusively on sight in this way, otherwise experience of a stereo soundstage, or sound-projection in electro-acoustic composition, would be a feeble affair.[50] The fact that we do not just hear a sound as coming from the same place where we see the dog barking does not mean that we have to see the dog in order to perceive the production of the sound. A deaf or blind person could perceive the dog making the

---

[48] It is interesting to note also that many sonorous qualities have connotations of touch and vision—rough and sharp, or bright and dark, for instance—as Lippman (1977: 51) discusses.

[49] What has become known as the McGurk effect is the same phenomenon; when subjects hear the sound 'ba' while seeing the lip movement for 'ga', they think that they are hearing 'da' (McGurk and MacDonald 1976).

[50] Compare Nudds's claim that our usual way of thinking of sounds is as of things that are distinct from their sources, with composer Robert Worby on the aesthetic consequences of sound reproduction. Worby argues that sounds are identified almost invariably with respect to their producers: '... we rarely have difficulty identifying the source of what we're hearing ... But our language for articulating this experience is extremely limited—certainly compared with that of visual phenomena ... We usually call [the sounds] trumpet, motorbike, telephone ... Our vocabulary relating to sound is incredibly impoverished' (Nudds 2001: 221; Worby 2004).

sound by putting their hand on its throat, and experiencing the production of the sound by touch. More important for present purposes, there is also a sense in which one can experience the production of sounds through hearing alone. The twofold account of the experience of music which I have argued for in this chapter—which qualifies or rejects the acousmatic thesis—suggests that this is so. Thus one should not simply distinguish between acousmatic experience through hearing, and non-acousmatic experience through hearing and other senses. To return to my comments on Strawson earlier, one needs to have other senses in order to develop objectivity-concepts, but those senses do not need to be exercised on a particular occasion in order for one to apply such concepts on that occasion. The distinction is made clear by Gareth Evans, commenting on Strawson's discussion of the purely auditory world:

We can think of sounds as perceptible phenomena, phenomena that are independent of us, because [unlike the subject in the sound world] we have the resources for thinking of the abiding stuff in whose changes the truth of the proposition that there is a sound can be regarded as consisting... what enables us to think of sounds as being drowned out, and in this way, existing though unperceived, is the knowledge that their categorical basis—the scrapings [of the violin]—continues... (Evans 1980: 104–5)

Thus, it is incorrect to claim that 'We never simply *hear* something as producing a sound because we can't hear the sources of sounds apart from hearing the sounds that they make'. I can simply hear the cello as producing the sound, since I can experience at least the apparent location of the sound—an objective matter—on the basis of hearing alone.

On the humanistic standpoint I have defended here, both acousmatic and non-acousmatic experience are essential to musical experience. But although we can have non-acousmatic experience through hearing alone, we also have non-acousmatic but genuinely musical experience of music through the other senses. Thus one should not necessarily agree with composers such as Witold Lutoslawski when they insist that their works belong to a purely auditory world, and that musical value resides in the 'world of sound'. Lutoslawski writes:

Some people are inclined to interpret music in an extra-musical way. The world of sound alone is not rich enough for them... music alone cannot encompass their idea of music. Less sensitive listeners feel alien in the world of sound; their thoughts escape to a realm of images or feelings that do not exist in a piece of music. This is a subjective reaction.... (Lutoslawski 1989: 9)

These comments are correct and understandable as far as they go. At first one might assume that Lutoslawski wants to reject the naïve popular view that musical experience is not purely auditory but involves images or feelings. On closer examination, however, it turns out that on a deeper view also, music is not purely auditory, but through its non-acousmatic aspect addresses other senses in the act of performance.

References

Adorno, T. (1997). *Aesthetic Theory*, R. Hullot-Kentor (trans.). London: Athlone.
Burkert, W. (1972). *Lore and Science in Ancient Pythagoreanism*, E. Minar (trans.). Cambridge, Mass.: Harvard University Press.
Chion, M. (1994). *Audio-Vision: Sound on Screen*, C. Gorbman (trans.). New York: Columbia University Press.
Clark, P. (2005). 'James Tenney—All Shook Up'. *The Wire*, 253: 38–43.
Cox, C. and Warner, D. (eds.) (2004). *Audio Culture: Readings in Modern Music*. London: Continuum.
Dack, J. (1994). 'Pierre Schaeffer and the Significance of Radiophonic Art'. *Contemporary Music Review*.
Davies, S. (2001). *Musical Works and Performances*. Oxford: Clarendon Press.
De Lorenzo, R. and Eilers, M. (1991). 'Lights and Siren: A Review of Emergency Vehicle Warning Systems'. *Annals of Emergency Medicine*, 20: 1331–5.
Dhomont, F. (1995). 'Acousmatic Update'. *Contact*, 8.2. Available at <http://cec.concordia.ca/contact/contact82Dhom.html>.
Emmerson, S. (ed.). (1986). *The Language of Electro-acoustic Music*. London: Macmillan.
—— (2000). *Music, Electronic Media and Culture*. Aldershot: Ashgate.
—— and Smalley, D. (2001). 'Electro-acoustic Music', in Sadie and Tyrrell (2001).
Evans, G. (1980). 'Things without the Mind', in Z. Van Straaten (ed.) (1980). *Philosophical Subjects: Essays Presented to P.F. Strawson*. Oxford: Clarendon Press.
Hamilton, A. (1999). 'The Aesthetics of Western Art Music' (Discussion of R. Scruton's *The Aesthetics of Music*). *Philosophical Books*, 40: 145–55.
—— (2003a). 'Spectral Music'. *The Wire*, 237: 42–9.
—— (2003b). 'Review of Liquid Architecture/Bernard Parmegiani'. *The Wire*, 235: 90.
—— (2007a). 'Music and the Aural Arts'. *British Journal of Aesthetics*, 47: 46–63.
—— (2007b). *Aesthetics and Music*. London: Continuum.
Harrison, J. (1999). 'Imaginary Space: Spaces in the Imagination'. *Proceedings of Australasian Computer Music Conference 1999*.
Harvey, J. (1999). *In Quest of Spirit*. Berkeley: University of California Press.
—— (2008). 'Buddhism and Music', in M. Paddison and I. Deliège (eds.) (2008). *Contemporary Music: Theoretical and Philosophical Perspectives*. Aldershot: Ashgate.
Heathcote, A. (2004). 'Liberating Sounds: Philosophical Perspectives on the Music and Writings of Helmut Lachenmann'. Durham: Durham University MA thesis.
Holmes, T. (2003). *Electronic and Experimental Music*. London: Routledge.

Kahn, D. (1999). *Noise Water Meat*. Cambridge, Mass.: MIT Press.
Lee, R. K. (1988). 'Fu Ho U vs. Do Re Mi: The Technology of Notation Systems and Implications of Change in the Shakuhachi Tradition of Japan'. *Asian Music*, 19(2): 71–81.
Levinson, J. (1991). 'The Concept of Music', in *Music, Art and Metaphysics: Essays in Philosophical Aesthetics*. Ithace, NY: Cornell University Press.
Lippman, E. (1977). *A Humanistic Philosophy of Music*. New York: New York University Press.
—— (1999). *The Philosophy and Aesthetics of Music*. Lincoln, Nebr.: University of Nebraska Press.
Lopez, F. (accessed 2004). Interview at <http://www.franciscolopez.net>.
Lutoslawski, W. (1989). 'Witold Lutoslawski in Interview'. *Tempo*, 170: 4–12.
McGurk, H., and MacDonald, J. (1976). 'Hearing Lips and Seeing Voices'. *Nature*, 264: 746–8.
Manning, P. (1993). *Electronic and Computer Music*, 2nd edn. Oxford: Oxford University Press.
Morgan, R. (1991). *Twentieth-Century Music*. New York: Norton.
Nettl, B. (2001). 'An Ethnomusicologist Contemplates Universals in Musical Sound and Musical Culture', in N. Wallin, B. Merker, and S. Brown (eds.) (2001). *The Origins of Music*. Cambridge, Mass.: MIT Press, 462–72.
Nudds, M. (2001). 'Experiencing the Production of Sounds'. *European Journal of Philosophy*, 9: 210–29.
Palombini, C. (accessed 2004). 'Musique Concrète Revisited'. <http://www.rem.ufpr.br/REMv4/vol4/arti-palombini.htm>.
Rosen, C. (1999). *The Romantic Generation*. London: Fontana.
Sadie, S., and Tyrrell, J. (eds.) (2001). *New Grove Dictionary of Music and Musicians*. New York: Oxford University Press.
Schaeffer, P. (1966). *Traité des objets musicaux*. Paris: Editions du Seuil.
—— (1987). Interview with Tim Hodgkinson. *Recommended Records Quarterly*, 2/1. Reprinted in D. Rothenberg and M. Ulvaeus (eds.), *The Book of Music and Nature*. Middletown, Conn.: Wesleyan University Press, 2001, 34–44.
Schafer, R. M. (1969). *The New Soundscape*. Scarborough, Ontario: Berandol Music Limited.
—— (1977). *The Tuning of the World*. Toronto: McClelland and Stewart.
Scruton, R. (1997). *The Aesthetics of Music*. Oxford: Clarendon Press.
Sethares, W. (1997). *Tuning, Timbre, Spectrum, Scale*. Berlin: Springer-Verlag.
Smalley, D. (1986). 'Spectro-morphology and Structuring Processes', in Emmerson (1986).
Spitzer, M. (2004). *Metaphor and Musical Thought*. Chicago: University of Chicago Press.
Stockhausen, K. (1989). *Stockhausen on Music: Lectures And Interviews*, compiled by Robin Maconie. London: Marion Boyars.
Strawson, P. (1959). *Individuals: An Essay in Descriptive Metaphysics*. London: Methuen.

Thomson, W. (accessed 2004). 'Musical Sound'. *Encyclopaedia Britannica*, <http://www.britannica.com>.

Van Leeuwen, T. (1999). *Speech, Music, Sound*. Basingstoke: Palgrave Macmillan.

Walton, K. (1988). 'What is Abstract about the Art of Music?' *Journal of Aesthetics and Art Criticism*, 46: 351–64.

Watson, D. (1991). *The Wordsworth Dictionary of Musical Quotations*. Edinburgh: Chambers.

Wilder, A. (1972). *American Popular Song*. Oxford: Oxford University Press.

Windsor, L. (2000). 'Through and Around the Acousmatic: The Interpretation of Electroacoustic Sounds', in Emmerson (2000), 7–35.

Wishart, T. (1986). 'Sound Symbols and Landscapes', in Emmerson (1986), 41–60.

——(1996). *On Sonic Art*, rev. edn. Amsterdam: Harwood Academic.

Withington, D. (1998). 'Siren Sounds: Do they Actually Contribute to Traffic Accidents?' *Impact*, Spring 1998, at <http://www.sound-alert.co.uk>.

Wollheim, R. (1980), 'Seeing-as, Seeing-in'. *Art and its Objects*, 2nd edn. Cambridge: Cambridge University Press, 205–26.

Worby, R. (2004). 'Loudspeakers Revolutionised the Way We Hear Music—But Not the Way We Talk About It.' *Guardian*, April 24, 2004; also at <http://www.guardian.co.uk/arts/guesteditors/story/0,14481,1201741,00.html>.

Zuckerkandl, V. (1969). *Sound and Symbol*. Princeton: Princeton University Press.

9

Speech Sounds and the Direct Meeting of Minds[1]

BARRY C. SMITH

Philosophers often claim that it is through speech that we make knowledge of our minds available to one another, and that it is through the medium of a shared language that we achieve a genuine meeting of minds. When combined with a conception of linguistic understanding as the direct perception of meaning in people's words, the view suggests that there is no barrier to knowing the minds of others. Certainly, when listening to a language we understand, we do not hear the acoustic speech signal as just a sequence of sounds: we hear what is being said. As a phenomenological observation, the claim is impeccable, but it is mistaken epistemologically to assume that when we hear words *as* meaningful it is because we hear meanings *in* the sounds, perceived as immediately present on the surface of speech. As I shall argue, talk of the surface of speech and the location of sounds is misplaced. What we directly perceive are the *sources* of sounds, and the source of speech sounds is the human voice. The experience of listening to speech gives us non-linguistic information about a voice as source, while the meanings we hear the voiced sounds to have are the meanings we as listeners have attached to those words: the meanings they have for us. Contrary to expectations, this inner model of linguistic understanding can still accommodate knowledge of what others are saying, and so presents no obstacle to our knowing what others have in mind.

---

[1] Versions of this chapter were given at the Second Dubrovnik Workshop in the Philosophy of Linguistics, in September 2006, and the Second Workshop in Philosophy of Language and Linguistics of the Irish Network of Philosophy of Language in University College Dublin in December 2006. I am grateful for comments from the audiences on both occasions, and for further discussion and comment to Ophelia Deroy, Jim Higginbotham, Guy Longworth, and Matthew Nudds, Georges Rey, Paul Pietroski, and Katerina Von Krikstein.

1. Introduction

Noise, or mere acoustic signals in the environment, can be heard as sounds in so far as there are agents of auditory perception. Perceivers can hear all manner of things: the sound of a taxi in the street, the sound of a door closing, of a dog barking, of a clock ticking, and so on. But there are also specific sounds that are heard not as the sound of objects, but as the sounds of subjects: most importantly, speech sounds. But not only these; groaning, laughing, and crying, like talking, all involve the voice in the production of what is a public display of a subject's inner experience. In what follows, I will concentrate on speech sounds and consider what enables us to hear them as the sounds of subjects, rather than objects. In particular, I shall argue that the basis on which an immediate connection between the mind of the speaker and listener is established is the experience of hearing *someone* talk rather than hearing *what they say*.

What enables us to hear some sounds as meaningful speech? The question is important, since philosophers have taken meaningful speech as the best guide to what another is thinking and the firmest evidence that they are thinking. But to understand the role that language understanding plays in the epistemology of mind, we must first give an account of the epistemology of understanding.

From an everyday perspective, language understanding is not problematic. People speak in a language we understand, and we immediately recognize what was said. It may not be obvious why they said it or what they mean in saying it, but we can recognize the words and sentences they uttered. In this respect, listening to speech in a language one understands is unlike hearing mere noise, although one can recreate that experience when listening to speakers of an utterly foreign language. In the foreign language case, we know people are saying something, or we think we do, though we cannot tell where one word begins or ends, or say which range of sounds makes up a complete phrase or sentence. As far as we are concerned, it sounds like babble, while to each of them their speech sounds are heard as sharing of intimacies, the idle talk of a moment, or requests for information. What distinguishes the two cases? Is it something in the sounds themselves, something in their reception by listeners, or some relation between the speaker and listener?

At first pass, it is easy to think that while listening to a foreign language in which one hears nothing of discernible significance, the speakers of that language are somehow adding meaning as an accompaniment to what is otherwise noise, whether anyone understands it or not. This gives the impression that the speaker merely emits noise in the hope that others will attach some significance to it. But neither the speaker nor her listener views things in just this way. The speaker does not take herself to be merely producing sounds: she feels her mind to be fully revealed and on show in the choice of meaningful words she puts into the public sphere. And to a listener who can understand her, she is not heard as first making sounds that subsequently need decoding. Rather, she is heard as saying things, imparting information, asking questions, making demands. And even listeners who cannot understand what she is saying take the speech sounds she emits to be more than mere noise. They may feel unable to pick up what is going on in her speech, but they notice that others can immediately react to it.

Despite the ease of talking to people in a familiar language a puzzle remains. Just how do we understand one another's speech so easily? How do we make the contents of our minds publicly available to one another through our talk? In listening to speech in a language we understand, we seem to have direct access to what someone is saying. But how is such knowledge possible from the mere fact of listening to particular sounds? How, from these unpromising materials, can we be put in touch with other people's intended meanings? The fact that, environmentally speaking, we are only presented with sounds is evident when reflecting on the case of an utterly foreign language. In such circumstances one hears, not words, but a continuous sound stream interrupted when the speaker pauses for an intake of breath. The difficulty becomes clear when we realize that the only publicly available evidence we have to go on in understanding one another is observable behavior. Strictly speaking, all we can show to one another is a sequence of sounds, gestures, and facial expressions. So how do we succeed in communicating something with a precise linguistic significance on the basis of these unpromising materials? How can the noises and movements we make convey something of semantic significance to others?

If we contemplate the surface of observable behavior in a way consistent with the intuition that all that can be shown there are sounds and movements, we will probably adopt a description of linguistic behavior cast in the restricted, physicalist terms W. V. Quine recommends. And if we then try to locate meaning among these observable facts, we will be forced to reconstruct the content of speech in terms of mere patterns in verbal behavior. The sounds speakers produce will be conceived as a response to a stimulus. The range of a speaker's assents and dissents to a sound in observable circumstances will fix the *stimulus meaning* for these vocables (Quine 1960: ch. 2). This drastically impoverished notion of linguistic meaning results from the attempt to meet a publicity of meaning requirement: the obligation to show how meaning can be publicly available to others as a matter of observed behavior in observable circumstances. The requirement must be met, according to Quine, since what people display in their behavior is all we have to go on in interpreting their utterances. However, the notion of stimulus meaning fails to square with the rich phenomenological experience we enjoy in listening to speakers talk in a language we understand. As noted already, we hear such speakers as saying things more determinate in meaning than is suggested by the limitations of stimulus meaning. And if we respect this latter intuition, and adhere to our everyday and common-sense experience of the meaningfulness of speech, we will need to find some other way to accommodate linguistically communicated meanings.

At this point, some will be tempted to locate meaning not on the surface of linguistic behavior but behind it, in the mind of the speaker. According to such a view, the speaker's intended meaning is hidden and becomes a matter of hypothesis for others. This new picture accepts, along with the previous one, that all we can show in speech behavior is a sequence of sounds and movements, and further accepts that these materials are insufficient to reconstruct the rich notion of linguistic content we are all familiar with as language users. It then attempts to save the fullness of linguistic meaning by locating it behind the surface of linguistic behavior in the mind of the speaker. The price to pay for respecting our commitment to determinate linguistic meanings is to abandon the publicity of meaning. But a view according to which 'assigning a meaning to an utterance by a speaker of one's language is forming a hypothesis about something concealed behind the surface of his linguistic behavior' (McDowell 1998b: 252) is unacceptable, according to John McDowell, Michael Dummett, and others, because it makes our understanding of one another 'a mere matter of guesswork as to how things are in a private sphere concealed behind their behavior'. And as McDowell points out, such a position distorts our immediate recognition of the meaningfulness of speech in a language we understand. As has been stressed already, in such cases hearers do not find themselves listening to uninterpreted sounds. McDowell's phenomenological insight is clearly right. Listening to speech in a language we understand is not a matter of first hearing noises and then going on to infer what they must signify. Rather, as McDowell says: 'Our attention is indeed drawn to ... something present in the words—something capable of being heard or seen in the words by those who understand the language' (1998a: 99). The content of others' speech is not hidden beneath the surface of overt behavior. We hear people not merely as producing sounds but as saying something. It is part of this phenomenological insight that we cannot turn what we hear people saying back into sounds.

McDowell now looks for a middle way between the two unacceptable positions just described, and attempts, on the basis of his insight about the phenomenology of understanding, to establish a credible epistemology of understanding and metaphysics of meaning that can accommodate the publicity thesis. He looks for:

[a] construal of the thesis that meaning can be fully overt in linguistic behavior: a construal according to which whenever someone who is competent in a language speaks, so long as he speaks correctly, audibly, and so forth, he makes knowledge of his meaning available—to an audience who understands the language he is speaking. (McDowell 1998a: 352–3)

According to McDowell, the two options first considered are wrong because they force us to choose between Behaviorism and Cartesianism about the mind. The mental is either reduced to patterns in behavior or retreats beneath the surface of behavior into an utterly private realm. But why should we take these to be the only options? We want our understanding of people's speech to engage with their inner lives, but at the same time we want what they say to be outwardly revealed to us. The dilemma we find ourselves in, according to McDowell, of being pulled in one direction or the other, is due to their sharing of a common assumption, which must be discharged to avoid the horns of the dilemma.[2] The common assumption is the thought that all people can outwardly present to us when they speak is a sequence of sounds and gestures; mere bits of behavior described in meaning-free terms. From there we seem forced to choose between finding meaning in meaning-free behavior—reducing meaning to patterns in otherwise uninterpreted verbal behavior—or to finding meaning preserved in the mind of the speaker, hidden from view as part of a private, inaccessible realm of inner items—what Quine called the museum myth. The way to discharge this assumption is to come to see that there is no sharp divide between inner and outer, between the intentional mind and outward behavior. By ceasing to dichotomize the mind and the body in terms of inner and outer, we leave room to find the mind fully exhibited in behavior rather than hidden behind it, screened off by uninterpreted sounds and movements. Just as the involvement of the mind in intentional actions goes right to the ends of our finger tips, so it reaches right out into the sounds we publicly articulate. The mind's involvement in action, linguistic and otherwise, does not stop short of the full outward display of intentional agency. That is why, from the perspective of the listener, 'the

---

[2] The dialectic here should be familiar to readers of McDowell, who would usually recognize these two unacceptable options as Scylla and Charybdis.

understanding of a language... consists in awareness of... unproblematically detectable facts' (McDowell 1998a: 331). '[T]he significance of utterances in a language must, in general, lie open to view, in publicly available facts about linguistic behaviour in its circumstances' (314). Otherwise understanding would consist in 'hypotheses about inner states of the speaker lying behind the behaviour' (331).

In many respects, McDowell is more Quinean than Cartesian. He regards Quine's commitment to the publicity of meaning as wholly admirable. Where Quine goes wrong, in McDowell's view, is in insisting that publicly observable behavior be characterized in meaning-free terms—in particular, in the terms the natural scientist would recognize. There is no need to insist on such limitations. Moreover, were we to adhere to such scientific—or, as McDowell and his followers like to say, scientistic—scruples, there would be no way to capture what takes place in linguistic behavior, no correct characterization of what we hear in one another's speech. We need to recognize the deliberate and purposeful behavior speakers give rise to for what it is—the intentional production of meaningful speech—and there is no way of doing so save by presupposing the meanings of the words whose use by a speaker we are describing. We must appeal to the significance these bits of behavior have in the linguistic practices of the speech community we belong to when reporting what members of that community are up to:

[S]hared membership in a linguistic community is not just a matter of matching in aspects of an exterior that we present to anyone whatsoever, but equips us to make our minds available to one another by confronting one another with a different exterior from that which we present to outsiders. (McDowell 1998b: 253)

The use of language cannot be rendered faithfully without presupposing the language in question in our description of that use. It is shared possession of a language that makes it possible for us to reveal the contents of our minds to one another on the surface of our speech behavior: '[a] linguistic community is conceived as bound together, not by a match in mere externals (facts available to just anyone) but by a capacity for a meeting of minds' (McDowell 1998b: 253). Shared command of a language equips us to know one another's meaning without needing to arrive at that knowledge by interpretation, because it equips us to hear someone else's meaning in his words.

Quine was wrong to think the surface of speech could be described in meaning-free, physicalistic terms. Such materials cannot support descriptions of what people are up to in acts of uttering the meaningful words and sentences we immediately take them to be uttering. According to McDowell's picture, meaning is no longer 'conceived as behind the surface of linguistic behavior but as residing on its surface' when that surface is located properly and not characterized in the shallow way Quine insists upon. The overt surface we display to one another can only be recognized when it is seen as activity characterized in normative and meaningful terms. When we encounter speech sounds made by members of our linguistic community, the meanings we hear in their words lie open to view on the surface of their practice: '... the outward aspect of linguistic behaviour is essentially content-involving, so that the mind's role in speech is, as it were, on the surface' (McDowell 1998a: 100).

What the phenomenological datum about hearing meaning in people's words is now meant to show us is how utterly misguided it would be, as part of the epistemology of understanding, to suppose that our minds engaged with a surface comprising anything less than meaningful speech. To recognize speech for what it is, its surface must be characterized richly in terms that show how meanings can be fully available to us in the experience of listening to one another: 'the senses of utterances are not hidden behind them, but lie open to view' (McDowell 1998a: 99).[3]

In what follows, I will point to overwhelming evidence that the rich texture of our linguistic experience in listening to speech cannot be found on the surface of that speech, but at this stage I am concerned with McDowell's reasons for thinking that it can.

Notice that, for McDowell, the richness of that surface, and what it makes available to us, is not available to just anyone. The outward aspect that matters can only be presented to those who understand the language: 'one hears more, in speech in a language, when one has learned the language' (McDowell 1998a: 333). Few, if any, of the linguistic features of that surface will be detectable by outsiders, as we can appreciate when listening to a foreign language. Whether one hears these sounds as meaningful or merely as noise, depends, we are told, on whether one possesses knowledge of the language. But it is the nature of the dependence of what we can hear on our knowledge of language that we need to be told more about. How does linguistic knowledge make the outward and significant aspect of speech available to us? How does it enable us to hear what is there on the surface? On these issues McDowell has little to offer,

---

[3] It may seem as if we could never be mistaken if we accept McDowell's view, but that is not his claim. We are fallible in our epistemology of understanding because we may think we are hearing meaningful speech, enjoying a genuine meeting of minds, hearing the meanings that are there of the surface, and yet be subject to an illusion or auditory hallucination. Nonetheless, either we are directly perceiving real speech or just getting counterfeit coin. What doesn't explain our fallibility is the idea that we are always engaging with something less than meaningful speech or less than full evidence of a mind on show, in linguistic behaviour from which we make at best risky inferences about what goes on at the real locus of mind and meaning.

and he is even less forthcoming on how we come to acquire the linguistic knowledge that gives us this capacity. He tells us the difficulty lies in having to answer the question: 'How can drilling in a behavioural repertoire [effecting a change in one's external behavior] stretch one's perceptual capacities—cause one to be directly aware of facts of which one would otherwise not have been aware?' (McDowell 1998a: 333).

One's natural inclination is to say that it can't, and that there is simply no answer to this question. It is the wrong question. Any plausible account of how we acquire the capacity to experience (certain) speech sounds as meaningful must begin elsewhere. We need to look at what linguistics and psychology tell us about the acquisition of language and the perception of speech.

2. Fodor versus McDowell on the Epistemology of Understanding

First of all, the phenomenological datum that we hear more in the speech sounds of a language we understand cannot by itself support McDowell's conclusions about the metaphysics and epistemology of speech. Further argument is needed. For the very same phenomenological insights are produced by Jerry Fodor to support the claims that speech perception must be the result of unconscious and automatic modular processes: 'You can't help hearing an utterance of a sentence (in a language you know) as an utterance of a sentence... You can't hear speech as noise *even if you would prefer to*' (Fodor 1983: 52–3). Thus: '"I couldn't help hearing what you said" is one of those clichés which, often enough, expresses a literal truth; and it is what is *said* that one can't help hearing, not just what is *uttered*' (55; emphasis in original). For Fodor: '... understanding an utterance involves establishing its analysis at several different levels of representation: phonetic, phonological, lexical, syntactic, and so forth' (64). This is the work of fast, dedicated, and mandatory cognitive processes that perform inference-like computations on their domain-specific representations. Here, Fodor cites William Marslen-Wilson and Lorraine Tyler's work on word-recognition, who tell us that:

... even when subjects are asked to focus their attention on the acoustic-phonetic properties of the input, they do not seem able to avoid identifying the words involved... This implies that the kind of processing operations observable in spoken-word recognition are mediated by automatic processes which are obligatorily applied... (Marslen-Wilson and Tyler 1981: 327; as quoted by Fodor 1983: 53)

The automatic and obligatory character of such processes, in contrast to the voluntary and reflective process of conscious deliberation, is the hallmark of the sub-personal. And this is the picture Fodor offers us of our response to speech sounds in a familiar language. Language comprehension is accomplished by means of an input module: an informationally encapsulated cognitive mechanism that responds selectively to certain informational inputs, and delivers its outputs to the central (thought) processes. The fast and automatic way in which we hear what is said, rather than merely appreciating the acoustic properties of the sounds, is evidence, for Fodor, of the workings of a sub-personal linguistic system whose products are delivered to consciousness but whose workings are cognitively inscrutable.[4] Thus, the phenomenological datum about what is consciously accessible when listening to speech in a language one understands settles nothing about the locus of linguistic significance, nor whether linguistic comprehension is a matter of direct perception or unconscious inference.

The experience of speech sounds is richer for those who understand the language than for those who don't. But it is also poorer, in that there is good evidence that the auditory processing of environmental noise and speech sounds may proceed in parallel; the result of experiencing speech sounds may, however, inhibit the auditory processing of non-speech sounds. To hear speech is not to listen to the sounds. But it is, as I shall argue, to listen to their sources: the voices of those who are producing them.

3. Phenomenology as Epistemology: Taking Experience at Face Value

What further considerations can McDowell offer for taking his observation about perceived speech not just as a phenomenological but as an epistemological claim? There are, I think, three considerations. First, we are invited to take the experience at face value, as our listening to the meanings that are present in people's words, for unless we view our experience of speech in this way—and assuming that we accept there is determinate meaning to what people say—there would be no knowing for sure what someone else meant

---

[4] Fodor stresses the way in which information about the acoustic properties of speech is lost when comprehension takes place. We know how Swedish and Chinese sounds, but do we know how English sounds? We fail to notice the absence of certain acoustic properties, as when phonemes have been spliced out of the middle of a recording of someone uttering a word, and yet those listening still hear the whole word (the so-called phoneme restoration effect).

in uttering a sentence. Secondly, there is no *encounter* with anything less than meaningful speech from which to construct a meaning for the sounds we hear people utter. Thus, hearing speech in a language we understand is a matter of direct confrontation with the meaningful surface of other people's linguistic behavior. Thirdly, there is no way to recognize the activity of speech for what it is without presupposing a command of the meanings of the words in the language spoken.

There is something to the second and third of McDowell's points, though they both need careful qualification. It is true that *we* do not encounter anything less than meaningful speech from which we assemble the meaning of people's utterances, but that is because our *speech processors* make contact with features at the *sub-personal* level. However, McDowell provides an argument in support of his second point to the effect that even if we could slow down the processing and examine what goes on, as we can in the case of our fast recognition of written words on the basis of the letters that compose them, there would be nothing corresponding to letters and rules from which we could assemble word meanings that could provide an explicit grounding for our understanding of them. I agree with McDowell that even if we do not go through explicit reasoning, and we suppose that meaning recognition occurs quickly and sub-personally, or as McDowell would say by means of a 'cognitive short-cut', there is no set of meaning-free items from which to assemble lexical meanings.[5] Like McDowell, I think we recognize word meanings as a whole and that this occurs at the conscious personal level and not in our sub-personal linguistic systems. But to say this is not to say that such word meanings are to be found in words present on the surface of someone's speech.

McDowell may also be right to say that we must presuppose the hearer's knowledge of the meaning of words in order to credit him with the capacity to recognize other people's linguistic behavior as meaningful speech. But, once again, there is no reason to suppose that the word meanings he has knowledge of are located externally in outward aspects of speech behavior, parts of a publicly shared language. I shall argue that meanings reside in the minds of speakers and hearers, and that the meanings we hear in people's words are the meanings we take their words to have. The words uttered, when recognized, are heard with the meanings they have for the hearer.[6] Thus, I have to reject

---

[5] See McDowell's (1998a: 117–18) argument by analogy with reading letters. In fact, the empirical evidence suggests a dual-route model of reading that involves both whole-word recognition and letter-by-letter spelling out. The data from language pathologies shows evidence of double dissociation, where patients can lose one capacity while retaining the other.

[6] Note that McDowell (1998a: 282), too, claims that 'command of a meaning is wholly a matter of how it is with someone's mind' and 'that a speaker means what she does...must be constituted at least

McDowell's first point, and show there is another way to secure knowledge of what someone else is saying that does not presuppose taking our experience of perceived speech at face value as an encounter with the external surface of linguistic behavior.

In seeking another route by which to secure knowledge of what someone else is saying, I am not simply engaged in a philosophical exercise of looking for alternative accounts of the epistemology of understanding. Nor is my alternative account solely motivated by qualms about the metaphysical extravagance of McDowell's picture, according to which meanings lie on the surface of speech episodes.[7] Rather, the motivation derives in part from the existence of conclusive empirical arguments against the possibility of locating linguistic properties in the sounds speakers produce. A credible philosophical picture of how we understand one another's meanings that is compatible with the best findings of the linguistic and speech sciences is better than one that is not. Thus, despite the subtlety, attractiveness, and phenomenological acuity of McDowell's view, there are many reasons to think it is wrong about the epistemology of understanding and wholly mistaken about the locus of linguistic significance. What is more, there are further phenomenological grounds for thinking that he overlooks the real basis for a meeting of minds. I will briefly state the empirical findings that put pressure on McDowell's view and offer further philosophical considerations against the account based on our phenomenological experience of sound. Let us look now at the empirical arguments.

4. How do we Come to Hear Words in the Sounds People Utter?

In hearing speech sounds, we are presented with a continuous sound stream with no gaps indicating the boundaries between words that we find in written language. If there were gaps between words in human speech, it would sound unnatural and hard to follow. Yet the fact that we confront a largely

---

in part by her physical and social environment'. But this is not just an externalist thesis about meanings: 'command of a word's meaning is a mental capacity...the mind [is] the locus of our manipulations of meanings...Meanings are in the mind but as [Putnam's] argument establishes, they cannot be in the head; therefore, we ought to conclude, the mind is not in the head' (276). Recognizing another's meanings is recognizing a bit of their mind, and because meanings as parts of the mind are literally out there, as part of the external environment we encounter, so too are these parts of their mind.

[7] McDowell tries to lessen our qualms about this rather magical sprinkling of meanings on the exterior surfaces of things in the world by describing such a world as 'enchanted'.

uninterrupted acoustic signal is not easy to reconcile with what we take ourselves to be hearing when listening to others speak. What we 'hear' is the articulation of discrete words and syllables that do not, strictly speaking, occur in the acoustic speech signal. In fact, much of what we supposedly 'hear' in the speech signal makes no public appearance at all. Word boundaries, non-overlapping syllables, restored phonemes; none of these items is present in the speech signal, and yet all are perceived as being there. Somehow the mind imposes such items on the sound stream presented to us. So, what are we listening to when we hear another speak, and how does auditory perception give us knowledge of it? I will claim that what we are listening to is the voice of the person talking. What we 'hear' that person as saying depends on processes that go beyond the information given.

Speech perception depends on a 'set of processes by which the listener extracts words from the continuous, rapidly changing, acoustic signal of speech'. In recognizing the sentence uttered from the continuous signal and 'the multidimensional properties of [the] acoustic stimulus, we have to analyze the frequency spectrum, identify phonetic features, segment phonological units', as well as initiate word recognition and deploy syntactic information. And we do all of this at lightning speed. On average, 'we perceive and produce about three words per second or one phone every tenth of a second' (Trout 2001, 2003). The difficulty of the task cannot be overestimated, 'because the acoustic realizations of a given word can vary greatly depending on speech rate, speaker's voice features, context, etc.' (Dehaene-Lambertz *et al.* 2005: 21). And yet, 'Despite their apparent variability, words, and the phonemes that constitute them, are...most often effortlessly identified' (ibid.).

Just how are perceptual constancy and categorical perception effects achieved when attending to 'a continuous, rapidly changing acoustic speech signal'? Is it by means of 'general auditory mechanisms or special speech decoding processes'? Are properties of phoneme perception essentially dependent on physiological properties of the auditory system, psychoacoustic mechanisms, or are they the upshot of domain-specific speech processors? The overwhelming evidence finds in favor of specialized speech processing mechanisms rather than just general auditory mechanisms.

In an experiment by Dehaene-Lambertz *et al.* (2005), subjects are presented with computer-generated sounds akin to speech sounds, which, after a while, subjects suddenly come to hear as syllables:

Many people exposed to sine wave analogues of speech first report hearing them as electronic glissando and, later, when they switch into a 'speech mode', hearing them as syllables. This perceptual switch modifies their discrimination abilities, enhancing perception of differences that cross phonemic boundaries while diminishing perception of differences within phonemic categories. (Dehaene-Lambertz et al. 2005: 21)

Different cortical regions are activated depending on whether the perceiver is in speech or non-speech mode. Event-related potential (ERP) and functional magnetic resonance imaging (fMRI) studies show that switching to the speech mode significantly enhanced activation of certain brain areas (left superior gyrus and sulcus) and were 'activated significantly more by a phonemic change than by an acoustic change' (with the same acoustic stimuli). These results and many more like them serve to 'demonstrate that phoneme perception in adults relies on a specific and highly efficient left-hemisphere network which can be activated in a top-down fashion when processing ambiguous speech/non-speech stimuli' (Dehaene-Lambertz et al. 2005: 21).[8] Such dedicated and, most likely, innate, speech-processing mechanisms make a significant contribution to the perception of speech. They are responsible for phoneme constancy, for our perceiving word and syllable boundaries, and much else. Clearly, such mechanisms go beyond the information given in their inputs. Not only is some information from the acoustic signal discarded in the process of chunking, ordering, and reducing the amount of auditory information we are exposed to, but crucial information can be added, as is shown in the phoneme restoration effect (Warren 1970). Phonemic representation is computed faster and more efficiently than corresponding acoustic representation of the *same* stimulus. The phonemic network, once activated by our speech processors, can have an inhibitory effect on the concurrent auditory representations to prevent interference from non-linguistically pertinent differences (Liberman et al. 1981).

So, for those who understand the language, the experience of speech sounds is richer than the experience of those who do not. But, in certain environmental ways, it is poorer, too. For although the auditory processing of an acoustic signal as sound and as speech may proceed in parallel, the result of experiencing sounds as speech may inhibit the auditory processing of non-speech sounds. This may be why when we listen to speech in a familiar language we do not listen to the sounds; but, as I shall argue below, we do listen to the source of the sounds: the voice of the person who is producing them.

Not only do speech-processing mechanisms have an effect on speech perception, but visual information from faces can also affect the auditory

---

[8] There are many other empirical arguments in favor of a specialized speech processor. See Dehaene-Lambertz et al. (2005) for review and references.

perception of phonemes. The powerful McGurk effect occurs when subjects listening to the sound /ba/ while seeing on film a face making the lip movements for /ga/ hear the sound /da/. What they 'perceive' is a blend of the audio and visual information (MacDonald and McGurk 1978). In addition, there is neuroimaging evidence of the interaction of cortical areas for voice and face recognition when listening to a familiar speaker (von Kriegstein *et al.* 2005). What normally sighted listeners take themselves to hear may always be an amalgam of information from different sources.

At the level of phonemes, the evidence that we simply pick up linguistic information from the environment is scanty. First, we do not detect discrete units like phonemes in the acoustic speech signal, so there is considerable rupture between features of the acoustic signal and what we perceive as being uttered. And yet, without the direct perception of phonemes, there cannot be direct perception of words made up from those phonemes, let alone the perception of word boundaries. The phenomenological experience we have when we hear speech cannot easily be reconciled with the empirical findings. The same perceived phonemes correspond to quite different acoustic properties, and the same acoustic properties correspond to differently perceived phonemes. And where a phoneme is deleted in the middle of a word and replaced by a cough, one will report hearing an utterance of the whole word, including the missing phoneme, with a cough in the background (Warren 1970). What all this shows is that perceived speech sounds do not correspond to actual surface features of the speaker's acoustic signal. Now it may be objected that this is because we are trying to locate the surface of speech in the wrong place; in what Wittgensteinians like McDowell call 'sub-bedrock' terms.[9] But what the empirical findings really show is the extreme difficulty of locating what we experience when we perceive speech sounds at the phonemic level in anything that could properly be regarded as a surface in any sense. And yet an alignment between what is perceived and what occurs out there on the exterior of speech is precisely what the direct realist account requires. Surely, it is more plausible that the supposed surface of speech is in fact a percept of hearers due to the information they bring to bear in the course of processing the auditory information they are

---

[9] When describing phenomena like speech and other human actions, Wittgenstein reminds us to recognize the ground that lies before us as the ground, and warns us not to dig below bedrock. The point is that below bedrock, justifications give out, and there is nothing to support the attribution of normative notions like meaning and intention at the level above: the level Wittgenstein is calling 'bedrock'. See Wittgenstein (1983: VI, 31) and McDowell's (1998b: 249–54) gloss on this.

given. We hear speech *as* the articulation of distinct phonemes making up words that contribute to a whole sentence. But to hear a sequence of sounds *as* the utterance of meaningful words and sentences is not the same as saying we hear the words *in* the sounds or hear meanings in words present on the surface of speech. The conception of a surface as McDowell describes it is more plausibly construed in terms of the phenomenological experience of hearers than as anything lying on the exterior surface of the speaker's behavior.

The moral is that 'linguistic information is projected by means of articulations but is not embodied in them'. Linguistic information is 'read into' rather than 'read off' these sounds. It is part of our 'specifically human way with sounds' to do so (Harris and Lindsay 2000: 203). The speaker may take himself to be going public on what he is thinking and see himself as putting his meanings right out there in his words, but however things strike him phenomenologically, he cannot succeed in putting more into the sounds he emits than they can actually bear. And in many cases, he simply cannot make the crucial linguistic properties appear publicly at all. All that is out there are sounds and marks. And it is language users like us, with the cognitive systems we have, that can make something of these linguistically caused items. From the perspective of the speaker experiencing himself as producing a rich string of meaningful words and phrases, he is like a person tapping out, or whistling, a tune for others to recognize. All he puts out there are some impoverished noises, but he hears the sounds he produces with the rich inner accompaniments that make what he is tapping to or whistling seem so obvious. To the listener trying to recognize the tune, it may sound like mere tapping or noise, much as speech sounds sound like noise to those listening to a foreign language. Those who know the language and identify words in the sounds uttered will hear the sounds with the meanings they have for them as a matter of their inner experience.

A correct view of language and our knowledge of language needs to account for our capacity to hear complex meaning in speech sounds and to produce sounds imbued with meanings in indefinitely many cases; it will have to explain our immediate readiness to produce and comprehend utterances of sentences we have never used or heard before. And it will have to explain how by these means we succeed in making our minds available to one another. However, the way knowledge of the language helps us to perceive more in the speech sounds of a familiar language is not by giving us an ability to directly perceive 'unproblematically available facts' that lie on the surface of speech. It is by bringing our linguistic knowledge to bear on auditory inputs in order to recover *more* than is given in the sound waves themselves. Much of this will be done by automatic and unconscious processes, though, as I acknowledged above, this is not where word meaning is to be found.

In effect, we have to compare two conceptions of language and knowledge of language:

(A) Speakers' knowledge latches onto properties of an external language.
(B) Speakers' knowledge determines the properties of their internally represented language.

The (A) conception of language, popular with philosophers of language and the folk, supposes that our competence is based on our acquiring knowledge of the observable facts of a public language. These facts are often supposed to be matters of convention we are taught and gradually adopt. The (B) conception of language, widely held by generative linguists, is that there is a largely innate basis for language in the brains of human language users, where language is now understood as the internal mechanism that enables us to speak and understand. According to this conception, many linguistic properties are due to the organization of our language faculties. Thus, the correct grammatical generalizations about our language—the ones we actually conform to—are neither consciously arrived at nor conventional regularities. They are the upshot of the workings of an internalized grammar.[10] Speakers have an innate capacity for language because of their native endowment with a universal grammar and their initial exposure to linguistic data: data that do not fully determine the language or (I-language) acquired, as the poverty of stimulus arguments tell us. The linguistic structures we deal with are internally generated in the mind of the speaker and assigned to sounds and marks which otherwise carry no linguistic information.

By contrast, McDowell thinks that what we perceive in speech, by virtue of having learned a language, is something lying open to view—the *surface* of linguistic practice. These are linguistic phenomena already there that we come to perceive as a result of acquiring knowledge of the language: a range of facts that were not previously (directly) perceptible come into view as we 'find our way into' the language.

---

[10] To appreciate how radically different the linguist's notion of a grammar and language are from the traditional folk conception embraced by many philosophers, consider these remarks by Chomsky: '... what should we take as a language...? The natural choice is g, the generative procedure; thus a person who knows language L has a specified method for interpreting arbitrary expressions, such as ["Who do you think that John saw" and "What do you wonder who saw", "Who do you wonder what saw", "He likes John", "His mother likes John", "John likes him", "John's mother likes him", J (a sentence of Japanese)]. Let us call g_E the I-language that some particular speaker of English (Jones) has acquired' (Chomsky 1987: 181).

5. Syntax and the Surface of Speech Behavior

McDowell's picture of the meaningful surface of speech faces even greater difficulties when we look at the syntactic structure of sentences. The semantic interpretations we can give to word strings depend on what syntactic analysis they are given. What we hear an uttered sentence as meaning depends systematically on its linguistic form. An ambiguous word string can be heard first one way and then another, depending on how we perceive it as structured:

(1) He talked to the woman from the sailing boat.

Connections between linguistic form and meaning are what compositional theories of meaning set out to describe:

> If (but only if) speakers of the language can understand certain sentences they have not previously encountered, as a result of acquaintance with their parts, the semanticist must state how the meaning of these sentences is a function of the meanings of those parts. (Evans 1975: 344)

And they can do so only if they identify the semantically relevant structural constituents of a sentence. This depends on the internal syntactic organization of the sentence. Syntactic configurations constrain the interpretations that can be given to word strings. In the following examples, there are certain interpretations they cannot have and others they must have:

(2) I know Mary expected to feed herself.
(3) I wonder who Mary expected to feed herself.

In (2), 'Mary' and 'herself' can only be construed as referring to the same person, while in (3) 'Mary' and 'herself' cannot be so construed despite the same sequence of words appearing in both (2) and (3). Speakers know these facts but they do not know how they know them. In particular they do not know that 'who' is construed as referentially dependent on a phonetically empty category in the syntax, PRO, that serves as the arbitrary subject of 'to feed herself'. What language users hear these sentences as meaning—and what they cannot hear them as meaning—are systematically correlated with facts about the syntactic configuration of these strings. But where should we locate these syntactic facts and why do the interpretations speakers and hearers give to particular word strings conform to linguistic generalizations defined over such facts? The linguistic generalizations speakers conform to cannot be captured in terms of surface properties of these strings—assuming for the sake of argument we could unproblematically recognize words in the surface sound string. They consist, rather, in facts about hierarchical relations among constituents of sentences, only some of which appear in the surface string. It is well known in linguistic theory that we cannot describe the syntactic structure of sentences by reference to linear arrangements of word strings, and that we must posit levels of syntactic structure remote from surface form.[11] The question for us is how does syntactic information impact what we are able to hear in listening to speech in a familiar language?

McDowell recognizes the importance of systematicity and states the requirement of *system* in a theory of meaning as follows: 'We want to see the content we attribute to foreign sayings as determined by the contribution of distinguishable parts or aspects of foreign utterances, each of which may occur, making the same contribution, in a multiplicity of utterances' (1998a: 145).

This will be achieved by constructing a truth theory for the language, whose axioms deal with the primitive expressions of the language and feature as premises in the derivation of T-theorems that deal with sentences in which those expressions occur. Does this requirement only apply to theories constructed to interpret foreign languages? After all: 'Comprehension of speech in a familiar language is a matter of unreflective perception, not bringing a theory to bear' (McDowell 1998a: 179).

However, theory has a role in the home case, too, as it provides a means of describing the range of our capacity to perceive speech in a familiar language: 'The ability to comprehend heard speech is an information-processing capacity, and the theory would describe it by articulating in detail the relation, which defines the capacity, between input information and output information' (179). The range of facts about sentences, as we see deduced in the theory's output theorems, amounts to a description of the extent of the speaker's capacity.

> For theorems to be so deducible, utterances must be identifiable in terms of structures and constituents assigned to them by a systematic syntax; and it must be possible to match up those structures (if necessarily obliquely, through transformations) with configurations observable in physical utterance-events. (McDowell 1998a: 145)

More surprisingly, we hear:

The hard physical facts, then, that constrain the construction of a truth-characterization for a language actually spoken are (i) the structural properties of physical

---

[11] For more on this point and on the significance of it for the folk view of language, see Smith (2006a).

utterance-events that permit the language to be given a syntactic description; and (ii) the complex relations between behaviour and the environment that permit (some of) the behaviour to be described and understood in intentional terms. (1998a: 146)

It is at the level of theorems dealing with the content of whole sentences that the truth-theory 'makes contact with the hard physical facts'. If the theorems are to be deducible in the systematic ways described, they 'must characterize utterances in terms of structures and constituents; so that the relation of match... must hold between the structures assigned to sentences by the syntax with which the theory operates... and configurations observable in the physical utterance-events'. So, the requirement of systematicity 'makes itself felt... in connection with the match between theoretical syntax and actual utterance-events' (McDowell 1998a: 146). But the talk of a match between syntactic structure and the hard physical facts about actual utterance-events is even less empirically plausible than the identification of phonemes and syllables in the acoustic speech signal.

The syntactic structures that feature in the linguistic generalizations speakers conform to are not consciously recognized or manipulated by speakers in producing and hearing meaningful utterances; but in order to conform to such generalizations, speakers must be able somehow to register the relevant facts about the underlying syntactic structure of a sentence. Thus, it is much more plausible to suppose that 'the language-input system specifies, for any utterance in its domain, its linguistic and maybe its logical form', and that 'the language processor delivers, for each input utterance, a representation which specifies its lexical constituents *inter alia*' (Fodor 1983: 90–1). The resultant understanding of the uttered sounds depends on speech processing in which a syntactic analysis is provided for the lexical items recovered from the input. Not all of the properties of sentences' linguistic or logical form are phenomenologically accessible, but what is accessible—our hearing a sentence as structured—depends crucially on what syntactic structure our fast, automatic, and unconscious speech processors assigns to the input string. Once again, the linguistic properties we rely upon in understanding the speech of others are not properties we find on the surface of speech. The syntactic constituents, their categories, and syntactic dependence—along with phonetically null, empty categories like traces and PRO—are simply not found in the sound string. The case for locating the meaning-determining properties of syntax in the sounds we encounter is without empirical foundation. Instead, what we see at work in McDowell's picture, as in the case of phonemes, is the myth of the externally given nature of language.

At this point, McDowell and others with an exteriorized conception of language could suppose there was a dichotomy between the phonological and syntactic properties of language that belong to the language faculty, and the publicly accessible properties of word and sentence meaning that are consciously accessible. While the former cannot be located on the surface of speech, perhaps the latter reside in the sounds speakers make. In effect, this would be to deny that there was a single locus of linguistic significance, and to suppose instead that the meaning properties of words were public and social, while the phonological and syntactic properties were part of our internal cognitive psychology. This hybrid picture would preserve the idea of word meanings occurring at the personal level, while most of the other linguistic properties were represented sub-personally in the language faculty.

How plausible is the hybrid picture? Prima facie, it faces severe difficulties in describing how the *meanings* of words and sentences come to be directly related to sounds, since the words and grammar they depend on cannot be located auditorily. A supporter of this picture would have to show how the properties that reside at these different levels and locations either interact or could be aligned so as to respect linguistic generalizations. I doubt whether this could be done, but I think that such a position faces greater difficulties still, and that the very notion of a surface to speech becomes problematic when we reflect on the nature of our auditory experience of sounds.

6. Sounds and the Phenomenological Experience of Speech

It is crucial to McDowell's picture that the unproblematic meaning facts we supposedly perceive in others' speech reside on the surface, or outward aspects, of linguistic behavior. According to this view, we are able to know what people say only because we perceive meaning in the sounds they utter. The idea of locating the surface of speech where linguistic meaning is to be found makes sense only if we can locate the speech sounds in which meaning is meant to reside. But can we? I shall now argue that there is no surface to speech because our auditory perception of speech fails to *locate* the sounds speakers produce.

Sounds, in general, are hard to place in the spatial world and auditory perception gives us no clues as to where they might occur. When we reflect on the metaphysics of sounds, there appear to be only three candidate locations. Sounds are (i) at their sources, (ii) with us when we hear them, or (iii) somewhere in between. None of these options is satisfactory.[12] Sounds cannot be where their sources are, since the source of a loud explosion may be hundreds of miles away when we hear it. On the other hand, to treat sounds as occurring where I am when I hear them is to suppose that different hearers cannot literally hear the same sound. A bell is struck and it chimes in many minds at once. Surely they hear the same sound? On the second view, this could not be true. The final view usually treats sounds as identical with the sound waves that travel from the source to my ears, which would require sounds to travel, to get nearer and nearer to us as hearers. But we simply do not hear sounds themselves as moving. Of course, we can hear the source of the sound as getting nearer or farther from us, but sounds themselves are not heard as traveling towards us.

So what should we say? The origin of the view of sounds and sound perception I want to endorse is found in the work of Brian O'Shaughnessy (Chapter 6; 2000) and is developed further by Matthew Nudds in this volume (Chapter 4). O'Shaughnessy begins with a version of the second thesis, saying that 'the sound that I hear is *where I am* when I hear it'. Whether or not this is the correct account of sounds, the reasons he gives for this view offer an important insight into the non-special nature of our perception of sounds:

[H]earing the sound to be coming from point p is not a case of hearing it to be *at p*. This is because the sound that I hear is *where I am* when I hear it. Yet this latter fact is liable to elude us because, while we have the auditory experience of hearing that a sound *comes from p*, we do not have any experience that it is here where it now sounds. (Rather, we work that one out.) And this is so for a very interesting reason: namely, that we absolutely never immediately perceive sounds to be *at* any place. (O'Shaughnessy 2000: 446; emphasis in original)

Although we do not hear sounds as located, we do hear their sources as located. In developing this view, Nudds (Chapter 4) points out that sounds need not exhaust the immediate objects of our auditory attention. We commonly listen to their sources, not to the sounds themselves. We seldom pay attention to the sensory qualities of sounds, but we focus instead on what is producing them. We mostly perceive sounds in terms of the objects that produced them. We hear the sound *of* a violin, the sound *of* a dog barking, the sound *of* the logs on the fire, the sound *of* the gas being lit. We hear and are interested in these distal sources of the sounds, and we hear them because the experience of a sound represents its source and the properties of its source. Through the processing

---

[12] For stout defenses of the first and second positions, see Casati and Dokic (Chapter 5) and O'Callaghan (Chapter 2).

of sound waves we are able to tell quite a lot about the size, movement, and density of the objects producing the sounds. It is these properties of the things producing the sounds—and not the sounds themselves—that we are interested in. Nudds (Chapter 4) stresses that 'we cannot explain why the auditory system groups the frequency components that it detects in the way it does, other than in terms of a process that functions to extract information about the objects [sources] that produced those frequency components' (p. 74), and thus, 'we can only explain why we experience the sounds we do in terms of a process that functions to tell us about the sources of sounds' (p. 75). 'Auditory perception tells us about the sources of sounds' (p. 72).

In hearing sounds, we listen to what (or who) is producing those sounds. And in the case of speech, we listen primarily to a voice. Voices are the sources of speech sounds, and voices are special. A voice belongs to a person, an embodied subject who intentionally produces the sounds we hear. We can recognize a lot about the producer of those sounds from properties of the voice, and we succeed in recognizing people's voices after only a brief exposure to their speech. On the radio, we identify a person speaking from his or her voice. Recognizing a voice is in normal circumstances recognizing who is speaking. In normal conditions, a voice provides a unique sensory print of a person.[13] Very specific information about an individual is conveyed by the voice. The identity of an individual is recognized by voice quality—recognition of a voice is usually recognition of a person. Even emotional states are largely recognized by non-semantic properties of speech, as demonstrated by several experiments that show vocal expression of emotions as being reliably recognized in content-masked speech signals (Fukuda and Kostov 1999; Scherer *et al.* 1972). Voice typically conveys information about the size, age, and gender of the speaker. When we do pay attention to the sound of someone's voice, it is because it can tell us something about the source of the sounds: the person himself and the state of mind he or she is in. We may attend to the tone of his voice or to its loudness, and we may hear its tremor or its catch. The auditory system detects these slight variations in voice quality—even in the sounds of unfamiliar voices—and registers them as signs of the nervousness or irritation in the speaker. We listen to such sounds when they tell us something specific about a person's mind. Perception of a voice as the source of speech sounds connects us immediately and intimately to the mind of another. There is a unique and direct meeting of minds, and all of this happens without semantic understanding. We sometimes hear the sound of voices talking in another room without hearing what is being said. But,

---

[13] I owe the idea of putting things this way to Anne-Lise Giraud.

again, the import of the experience is not just that there are sounds I am experiencing. I am hearing the voices of *people* talking. Through hearing a voice, we hear ourselves being addressed by a person, and, if the experience is veridical, we are hearing the mind of an individual who is addressing us. None of this information is conveyed via the content of what the person is saying. We do typically hear certain speech sounds *as* meaningful. But do we literally hear the meaning *in* the sounds? How can we say this on the basis of perceptual experience? The experience of speech sounds locates their sources (or apparent sources), but not the sounds.[14] We treat the source—a voice and therefore a subject—as the originator of the meanings we take the uttered words to have, but we do not perceive the meanings to be anywhere.

We do not have to claim that speech sounds do not have a location, though perhaps they do not. Rather, we only need the phenomenological observation that our experience of speech sounds fails to locate them. After all, where do you hear the sounds of someone's speech to be occurring? Look at a speaker's mouth moving and note what you hear the speaker saying. *Where* in this sequence do you observe the speech sounds to be occurring? It is impossible on the basis of our phenomenological experience of speech to give any location to the speech sounds in which meanings are meant to reside, we cannot give any external location to the surface of speech. The location is neither on the speaker's lips, where I am looking when I hear someone speak, nor in the air between us. (I hear the sounds as coming from that person—in other words, my experience locates the source of the sounds.) Without a place for speech sounds to be, there is no exterior surface on which to locate the meanings of words. Auditory speech perception simply gives us experience of sounds that presents voices and properties of those voices. One hears this as a result of auditory speech processing involving the segmenting and grouping of sound waves, with perhaps some knowledge of the properties of the human voice attended to. We do not experience the meaning of words as lying 'open to view, in publicly available facts about linguistic behavior in its circumstances', or as occurring anywhere. The phenomenological experience is of listening to a voice, the voice of a person. And what we take the words we recognize being uttered to mean is what we take the person who is voicing their thoughts to mean.

---

[14] We need to say 'apparent sources', because of the ventriloquism effect made use of in cinema, where an unlocalized source of a voice sound comes to be identified with the location of a visual cue of a mouth making speech sounds. Attentional capture of the auditory system by visual cues happens only so long as the lip synchronization with the sounds is close enough, otherwise the illusion breaks down. I am grateful to Charles Spence for this point.

The conviction we have that we are in touch with someone when talking to them, that there is no barrier between minds, is largely due to features of face and voice we recognize independently of understanding the content of their speech. The recognition of a voice, is, normally, the recognition of a person.[15] There are dedicated neural areas in the superior temporal sulci (STS) that respond selectively to voices more than to other sounds in the environment (Belin *et al.* 2000).[16] In particular, the anterior area is dedicated to voice processing and not the linguistic analysis of speech sounds (von Kriegstein *et al.* 2003). One reason why the recognition of a speaker happens so quickly is that it involves such early processing areas in the brain, such as the fusiform gyrus for face recognition and regions of STS for voice. Activation of these cortical regions helps us to quickly identify and form the capacity to recognize a person, and there is evidence from neuroimaging of the interaction of face and voice areas in the recognition of a speaker (von Kriegstein *et al.* 2005). Cross-modal integration occurs where we focus our attention visually on a speaker we want to listen to. We have to direct our visual attention in order to enhance our hearing of a particular person speaking at the other end of a table. In a crowded room or restaurant where many people are speaking at once, we need to direct our visual attention to orient our auditory perception to the *person* speaking as the source of the sounds we want to hear.

All of this shows that even without locating the content of speech sounds in auditory perception, there can still be a direct meeting of minds, due to our awareness of an individual subject or person as the *source* of the speech sounds we are hearing. It is natural to take the subject to be the originator of the meanings we attach to the words we retrieve from her acoustic sound stream in the course of lexical processing.

I hear you as saying something. But what I hear you as saying is the result of the meanings the words you utter have for me. I can only hear your words with the meanings I attach to them. Who else's meanings would I use, other than my own? Thus, if a word (a set of phonemes) has certain meaning for you that it does not have for me, I can only hear it with my meaning, not with yours. Similarly, if a word is ambiguous for me but not for you, you simply cannot *hear* it with one of the meanings I give it, even if you can come to know it has that other meaning for me. Thus, if a shop assistant in Glasgow says, 'Would

---

[15] Although voice is a property of a person, a person can have more than one voice. We talk of someone's 'singing voice' and are surprised by it even though we know their 'speaking voice'.

[16] STS is the area where we find mirror neurons that resonate to observed actions of others, perhaps suggesting that we may be able to find neural evidence for the motor theory of speech perception, according to which we are helped to perceive sounds by our motor system's matching of the articulatory movements that produce them.

you like a wee poke?' I may hear the question as meaning, 'Would you like a paper bag?' while you may not.

7. How Do I Come to Know What Others are Saying?

Through the early learning of word meanings, children come—in contexts involving another language user and under conditions of joint attention—to attach a meaning to a word or sound they hear. It may appear as though the child is being given the meaning of that word, but from the child's point of view, it is learning to endow that sequence of sounds with a meaning. And it is the meanings speakers have endowed their words with that count as their default understanding of these words whenever they encounter them. This is what they hear the words as meaning. Of course, the default case can be overridden and one can be wrong to take this to be what someone else means by their use of these words. But it will fix, initially, what we *hear* them as saying. And this will be a matter not of detecting meanings in their overt speech, but in our contributing the meanings we usually attach to the words we perceive to our understanding of their speech. When these are also the meanings they attach to those words, we will count as knowing what they mean, as being correct in what we take them to be saying.[17]

In listening to your voice, I am directly in contact with you as a person; but in hearing you say certain things, I supply meanings for the words I recognize you to be uttering. I simply always experience these words, at first, as said or heard with the meanings they have for me—the meanings I have endowed them with. The immediacy of the experience I have in hearing what you say is due to the inseparability for me of these words and these meanings. If my immediate understanding of you does not work, and the default condition—where you and I have attached the same meanings to these words—fails, I need to distance myself from my immediate understanding and engage in interpretation.

Notice how this picture differs from the one McDowell seeks to resist. He told us that 'the significance of utterances in a language must, in general, lie open to view, in publicly available facts about linguistic behavior in its circumstances' (McDowell 1998a: 314). Otherwise, understanding would

---

[17] For an account of how we first attach meaning to words and use these in a first-person-based epistemology of understanding, see Smith (2006b, 2006c).

consist in 'hypotheses about inner states of the speaker lying behind the behavior' (McDowell ibid.: 314, 331). But on my picture, where meanings do not 'lie open to view' on the surface of linguistic behavior, we are not as listeners hypothesizing about others' inner states. We are just hearing the words retrieved from the speaker's speech signal with the meanings they must have for us. Initially, we have no choice but to hear them this way. Our task is not to infer what goes on with others, but just to hear them as we are naturally and immediately inclined to do. By default, what we take someone to be saying will be what they are saying. According to this view, the direct connection with the mind of another will occur via perception of the *source*, and not the *content* of the speech. It is the sound of a person, not what the person says, that establishes a meeting of minds.

I can be mistaken about what you are saying, but if you are addressing me, I will not be mistaken about its being *you*—my interlocutor—who seems to be saying these things. But can we not make mistakes about who or what is addressing us in speech? Not if the experience is veridical, I say. But what is required for our experience of speech to be veridical? Nudds points out that 'the experience of sounds commits us to the existence of something other than sounds'. The experience of a violin being played is veridical if there is a violin being played and it is the source of the sounds heard.

Sound waves carry information about the things that produce them, and, thus, we can perceive those things through the auditory experiences that represent those objects and their properties. The auditory speech system functions to produce experiences of hearing a voice, and having a voice is a property of a person. In the auditory perception of speech, we hear the speech sounds as coming from a person who is speaking to us. So, when it is veridical, our experience of speech sounds commits us to the existence of voices which belong to persons. The correctness conditions for auditory experiences of speech sounds have these existential commitments, but they do not carry commitments about meanings residing on the surface of speech or lying open to view. Since the experience of sounds and their sources does not commit us to a surface for speech. Our experience of sounds does not provide location for those sounds, only for their sources. Usually, we know who is producing these sounds, or think we do. And yet, speech synthesizers produce powerful illusions as if we were encountering a person with a personality. We hear meaning in what these faux voices say, in the usual way, but there is a strong pull to misperceive the source of the sounds as a person. People frequently describe the 'voice' of a speech synthesizer in automatic telephone or satellite navigation systems as sounding insistent or strident or cold. These are abnormal cases. Normally, we hear *someone* saying such and such, and that person is perceived as being the source, or apparent source, of the sounds.

Sounds from speakers in our immediate community are heard *as* meaningful but the linguistic meanings and forms on which perceived meaning depends are not there in the sounds we hear. The internal organization of language users provides all the linguistic significance there is.[18] The real object of speech perception is the *voice* of the producer. We hear the minds of others in the sounds they make but not what they say. The sounds do not carry meaning; they trigger the awareness of meaning in us. Producing meaningful speech sounds is like tapping out a tune for others to catch on to, and those who have learned the same tunes may hear the sounds in the same way. All the richness we hear in meaningful speech is not in the sounds but in us.

References

Belin, P., Zatorre, R. J., Lafaille, P., Ahad, P., and Pike, B. (2000). 'Voice Selective Areas in Human Auditory Cortex'. *Nature*, 403: 309–12.

Chomsky, N. (1987). 'Reply'. *Mind and Language*, 2: 178–97.

Dehaene-Lambertz, G., Pallier, C., Serniiclaes, W., Sprenger-Charolles, L., Jobert, A., and Dehaene, S. (2005). 'Neural Correlates of Switching from Auditory to Speech Perception'. *NeuroImage*, 24: 21–33.

Evans, G. (1975). 'Identity and Predication'. *Journal of Philosophy*, 72: 343–63.

Fodor, J. A. (1983). *The Modularity of Mind*. Cambridge, Mass.: MIT Press.

Fukuda, S. and Kostov, V. (1999). 'Extracting Emotion from Voice'. *Systems, Man, and Cybernetics, IEEE SMC '99 Conference Proceedings*, 4: 299–304.

Harris, J. and Lindsay, G. (2000). 'Vowel Patterns in Mind and Sound', in N. Burton-Roberts, P. Carr, G. J. Docherty (eds.), *Phonological Knowledge: Conceptual and Empirical Issues*. Oxford: Oxford University Press, 185–206.

Liberman, A. M., Isenberg, D., and Rakerd, B. (1981). 'Duplex Perception of Cues for Stop Consonants: Evidence for a Phonetic Mode'. *Perception and Psychophysics*, 30(2): 133–43.

MacDonald, J. and McGurk, H. (1978). 'Visual Influences on Speech Perception Processes'. *Perception and Psychophysics*, 24: 253–7.

McDowell, J. (1998a). *Meaning, Knowledge and Reality*. Cambridge, Mass.: Harvard University Press.

——(1998b). *Mind, Value and Reality*. Cambridge, Mass.: Harvard University Press.

Marslen-Wilson, W. and Tyler, L. (1981). 'Central Processes in Speech Understanding'. *Philosophical Transactions of the Royal Society*, 8: 317–22.

---

[18] Notice, however, that although some of the objects of our linguistic knowledge are mental and internal, they depend for their content on others and on aspects of the environment.

O'Shaughnessy, B. (2000). *Consciousness and the World*. Oxford: Clarendon Press.
Quine, W. V. (1960). *Word and Object*. Cambridge, Mass.: MIT Press.
Scherer, K. R., Koivumaki, J., and Rosenthal, R. (1972). 'Minimal Cues in the Vocal Communication of Affect: Judging Emotions from Content-Masked Speech'. *Journal of Psycholinguistic Research*, 1: 269–85.
Smith, B. C. (2006a). 'What I Know When I Know a Language', in E. Lepore and B. C. Smith (eds.), *The Oxford Handbook of Philosophy of Language*. Oxford: Oxford University Press.
—— (2006b). 'Davidson, Interpretation and First-Person Constraints on Meaning'. *International Journal of Philosophical Studies*, 14(3): 385–406.
—— (2006c). 'Publicity, Externalism and Inner States', in T. Marvan (ed.), *What Determines Content? The Internalism/Externalism Dispute*. Cambridge, Mass.: Cambridge Scholars Press.
Trout, J. D. (2001). 'The Biological Basis for Speech: What to Infer from Talking to the Animals'. *Psychological Review*, 108: 523–49.
—— (2003). 'Biological Specialization for Speech: What Can the Animals Tell Us?' *Current Directions in Psychological Science*, 12(5): 155–9.
von Kriegstein, K., Eger, E., and Kleinschmidt, A. (2003). 'Modulation of Neural Responses to Speech by Direction Attention to Voice or Verbal Content'. *Cognitive Brain Research*, 17: 48–55.
—— Sterzer, P., and Giraud, A. (2005). 'Interaction of Face and Voice Areas During Speaker Recognition'. *Journal of Cognitive Neuroscience*, 17(3): 367–76.
Warren, R. (1970). 'Perceptual Restoration of Missing Speech Sounds'. *Science*, 167: 392–3.
Wittgenstein, L. (1983). *Remarks on the Foundations of Mathematics*, rev. edn. Cambridge, Mass.: MIT Press.

10

The Motor Theory of Speech Perception[1]

CHRISTOPHER MOLE

There is a long-standing project in psychology the goal of which is to explain our ability to perceive speech. The project is motivated by evidence that seems to indicate that the cognitive processing to which speech sounds are subjected is somehow different from the normal processing employed in hearing. The Motor Theory of speech perception was proposed in the 1960s as an attempt to explain this specialness. It is currently enjoying a renewal of interest, partly on account of our developing understanding of mirror-neurons (the existence of which is suggestive but not conclusive) and partly on account of some recent work using Transcranial Magnetic Stimulation (Fadiga *et al.* 2002).

This essay has two parts. The first is concerned with the Motor Theory's explanandum and shows that it is rather hard to give a precise account of what the Motor Theory is a theory *of*. The second part of the essay identifies problems with the explanans: There are difficulties in finding a plausible account of what the content of the Motor Theory is supposed to be. The agenda of both parts is rather negative, and problems will be uncovered rather than solved. In the concluding section, I shall suggest where one might look if one wants to solve the Motor Theory's problems, but it is unclear whether the Motor Theory's problems *ought* to be solved, or whether the whole theory should be abandoned.

I.

Psychologists were first persuaded that speech perception is unlike the perception of other sounds by the failure of attempts to build reading machines

---

[1] This work was done while the author held the William Alexander Fleet Fellowship. Thanks are due to Ms Julia Fleet, who sadly died while this volume was in preparation.

for the blind. Nowadays our computers do a good job of rendering a written text into speech, but it was not always so easy. After the Second World War there were lots of recently blinded people, for whom a machine that could read aloud would have been a very good thing. There was also rather little computing power available. The task of building a machine that would turn text into *speech* seemed to be a practically impossible one, but the task of building a machine that would make *some* sort of distinct *noise* for each of the letters in a text seemed straightforward enough. The more ambitious task of building a machine that would make a distinct noise for each of the separate speech sounds in the text (that is, each of the *phonemes*) also looked like a real possibility.

Such a task may have been computationally tractable, but as a substitute for reading it was completely hopeless. The thing that made the project hopeless was that the listener's ears just couldn't keep up. If one's reading machine was making its sounds at a pace that was anything like the pace at which our mouths make sounds when we speak, then it was making sounds at a rate that was far too fast for the listener to resolve. If the sounds were given slowly enough for the listener to resolve them, then they came far too slowly to effectively communicate a text. Whichever sounds are allocated to individual letters or phonemes, the resulting auditory presentation of words takes much longer to comprehend and puts a much greater load on working memory than written text or speech. Training does little to help. As Alvin Liberman, one of the first and most prominent researchers on speech, puts it, 'Only the sounds of speech are efficient vehicles for phonetic structures; no other sounds, no matter how artfully contrived, will work better than about one tenth as well' (Liberman 1990).

The blind war veterans stood no chance at all of learning to recognize the rapidly presented sounds of the 'reading machines', but they had, like the rest of us, learned in infancy how to recognize the phonemes that make up speech, and these phonemes can be resolved at a very fast rate indeed. Why, then, could they not learn to resolve any other sounds at anything approaching the same pace? It seems that *somehow* speech must be special. Perhaps its specialness has something to do with its being learned so early on. Perhaps it is special because it is massively more familiar than other sounds, and massively more practiced, but perhaps it is special in other ways, too. Language so often is.[2]

---

[2] Not everyone in the psychology of speech perception has signed on to the 'speech is special' view, but dissent is certainly rare. The dissenting view can be found in the discussion of Miller's Auditory-Perceptual Pointer Model in Klatt (1989). Or Fowler's 'Direct Realist' theory (Fowler 1986).

To be told that speech is special is to be told *something* of interest, but it is not yet to be told anything sufficiently precise for us to be in a position to assess a scientific theory that purports to explain this specialness. Researchers (most of them based at Haskins Laboratories where the work on reading machines had been carried out) uncovered multiple effects that were thought to illuminate the *way* in which speech is special. Four, in particular, attracted the attention of psychologists more widely. These were: the categorical perception of speech, the lack of invariance, the 'duplex' perception of speech, and the McGurk effect. I'll explain each of these in turn.

Categorical Perception

To understand the categorical nature of speech perception, consider the discrimination function for normal non-speech sounds—the function that gives, for each magnitude of a perceivable property, the smallest discernible difference from that magnitude. By presenting normal listeners with pairs of sounds, and asking whether the two sounds sound the same or different, we can work out, for any variation in magnitude on any dimension, what the smallest discernible difference in that magnitude is. And with this data, we can plot a graph showing, for each of the pitches we have tested, how much deviation is needed from that pitch for the two sounds to sound different, or we can plot a graph showing how much change in *tone* is needed for two sounds to sound different, and so on for various other dimensions along which acoustic properties vary. These graphs showing the discrimination function for simple acoustic properties will tend to be gently curving lines. As we test a subject's ability to discern variations from louder and louder sounds, we don't come across any one particular volume from which deviations are much more easily detected than are deviations from slightly louder, or slightly quieter sounds. The ability to discriminate one pitch from another changes smoothly as a function of the initial pitch. The same is true for volume and timbre. This *isn't* true if the dimension of variation is a dimension that is relevant to differences in speech sound. If the sound we are testing is a speech sound—such as the syllable /pa/—and if the magnitude we are varying is one that can make a difference to which phonemes are heard—such as the voice onset time—then there *is* a privileged magnitude such that deviations from that magnitude are much easier to detect than deviations from the previous and from the subsequent magnitudes. The graph on which we plot the smallest discernible differences will not curve smoothly. There will be a sharp trough in it at the point where we suddenly become extremely sensitive to small variations. It will be helpful to spell out this example in some detail.

The syllables /pa/ and /ba/ both begin with bilabial plosive consonants. That is to say that the initial sound in both is produced by closing the lips, allowing a little air to build up behind them, and then releasing that air. They differ in the moment during this performance at which the vocal chords begin to vibrate. For a /ba/ the vocal chords vibrate almost as soon as the air is released. A /pa/ is produced when the vocal chords vibrate a little later. The time between the release of the air held at the lips and the beginning of the vocal chord vibration is called voice-onset time (VOT). VOT can be varied continuously and a spectrum of phonemes can be artificially produced, each of which differs from the preceding phoneme only by, say, an increase in VOT of 10 ms.

If we listen to each of the sounds in this continuously varying spectrum they are not *heard* as varying continuously. We don't hear a sequence of /ba/s gently sloping off into penumbral cases which gradually become recognizable as /pa/s. Instead, the first half of the spectrum is all heard as more or less the same /ba/ sound, and the second half of the spectrum is all heard as more or less the same /pa/ sound, and there is a narrow band in the middle of the spectrum, when VOT is around 26.8 ms, where subjects differ as to how they hear the sound. At this transition point subjects can recognize very slight variations in VOT, since these slight variations are enough to move the sound from one category to the next. From the point of view of the subject, the difference between a syllable with VOT of 26.8 ms and a syllable with VOT of 36.8 ms sounds like a really big difference, whereas a difference of the same objective magnitude, between, say, VOT of 30 ms and of 40 ms, sounds like a very slight difference. In the first pair, one of the syllables sounds like a /ba/ and the other sounds like a /pa/. In the second pair, the two syllables are more or less indistinguishable /ba/s.[3] This distinctive discrimination function is found for many of the dimensions of variation that, like VOT, make the difference between two consonants, although it isn't found for vowels.

This finding by itself is rather unremarkable. It isn't surprising to find a difference between the discrimination functions for complex sorts of variation, such as variation in VOT, and the discrimination functions for simple sorts of variation, such as variation in pitch. And it isn't surprising that we exploit

---

[3] This is a slight simplification of the experimental procedure used, but not, I think, a significant one. The usual procedure is not pair-wise comparison but comparison of triples. Subjects are presented with a pair of neighboring sounds—*A* and *B*—and then presented a third sound—*X*—which is just the same as either *A* or *B*. Their task is to discern which of the two initial sounds *X* repeats. With a VOT continuum where neighbouring items differ by 10 ms, subjects perform close to chance on the *ABX* task for almost all of the spectrum except for the point (when VOT is around 26.8 ms) at which *A* is heard as a /ba/ and *B* as a /pa/.

the categorical perception of such variations in the boundaries that we use to indicate semantically relevant differences in speech. Something similar to the categorical perception of VOT holds for the perception of colors: we perceive two reds as more similar than a red and a yellow, even though the objective difference between the wavelengths may be the same. If you were designing a communication system with colored flags, you'd assign different meanings to red and yellow, instead of making the difference between two reds a semantically significant one. The categorical perception of speech sounds might be used in the speech code in something like this way, without the fact that the phonemic boundaries coincide with the boundaries of categorical perception showing any connection between categorical perception and the specialness of speech perception.

Categorical perception might be a feature of the perception of complex variations—a feature that the speech code makes use of, but that is not a special feature of speech perception as such. Two things suggest this. First, it is found that there is categorical perception of some non-speech sounds, as when, for example, musicians show categorical perception for semitone boundaries, or when normal people show unlearned categorical perception effects for certain 'buzz, noise, and relative timing continua' (Harnad 1987: 9). Secondly, and more impressively, categorical perception is found in some creatures that lack language (and lack anything that might be thought of as a proto-language). The categorical perception of speech sounds was found in chinchillas by Kuhl and Miller (1978), in Japanese quail by Kluender *et al.* (1987), and in the Mongolian gerbil by Sinnott and Mosteller (2001).

This is not to say that categorical perception doesn't *contribute* to the specialness of speech, only that the categorical nature of speech perception isn't by itself adequate for characterizing the way in which speech is special. There is, in fact, good evidence that categorical perception *is related* to the specialness of speech coming from the fact that the categorical perception of speech is affected by the role it plays in language. We know that it is affected in this way because we know that the pattern of categorical perception that people show for speech sounds is affected by the pattern of phonemic contrasts that can make for a difference in meaning in their native language. English speakers, for whom the difference between /l/ and /r/ can be a semantically relevant difference, perceive a continuous change from one to the other as a categorical change. Their discrimination function has the trough that is characteristic of categorical perception. In Japanese, the difference between /l/ and /r/ is not a difference that ever distinguishes two phonemes. (That is to say that the difference between whether an /l/ or an /r/ was said never makes for a difference in which words the speaker uttered, unlike in English, when it makes for the difference between the request for a room, or for a loom.) Japanese speakers *don't* hear the change from /l/ to /r/ categorically (Miyawaki et al. 1975). This effect has been replicated for several other languages that have or lack specific phonemic contrasts. Moreover, it does not seem that these phonemic categories are learned. It is the *lack* of categorical perception for the phonemic differences not used in one's native language that gets acquired. Infants can perceive all the category differences used in any language, but they lose the ability to discriminate the differences not used to distinguish phonemes in their native language (Best and McRoberts 2003).

The categorical perception of speech points to a feature of the perception of complex sounds—a feature that has a role in the way speech is perceived, but not, it seems, a feature that is adequate, by itself, to account for the specialness of speech perception. Nor is it at all clear how crucial the role of categorical perception is. Really tidy categorical discrimination functions are only found in experiments where the subjects are preselected 'to insure that all subjects ultimately serving in the experiment would have a sharp and clear phoneme boundary' (Lane 1965). Those subjects who show categorical perception to a lesser extent may be at a disadvantage as language-learning infants (Espy *et al.* 2004), but there is no suggestion that they find words any harder to *hear*.

The Lack of Invariance

The second feature of speech perception that motivated the proposal of the Motor Theory is the 'lack of invariance' in the speech signal. In the literature on the Motor Theory, the lack of invariance is the most focused on feature of speech perception, and many writers treat lack of invariance as the theory's sole explanandum. Ivry and Justus (2001), for example, write, 'This theory [the Motor Theory] was proposed to account for our ability to perceive the invariant articulatory events that form the speech stream, in spite of the great variability of the acoustic signal' (Ivry and Justus 2001: 513). The emphasis on invariance is understandable, since it is the lack of invariance that makes it so hard to write reliable speech recognition software (and many researchers have been engaged in just that project), but it is unfortunate. A careful consideration of what lack of invariance *is* reveals it to be a feature that is shared by many perceptual tasks, not a special feature of speech at all. The speech signal lacks invariance in the following way: for any given phoneme there are various different patterns of acoustic energy that give rise to the experience of that phoneme being uttered, and for any given pattern of acoustic energy there are various different phonemes that can be experienced when that pattern of energy is heard. There is no mapping from simple features of the acoustic signal to phonemes heard.

This lack of invariance was revealed by the examination of speech spectrograms, an example of which is given in Figure 10.1. The dark patches on the diagram correspond to the frequencies at which there was acoustic energy. The *x*-axis represents time. The speech signal is invariant in that there is no pattern of acoustic energy—no pattern of dark patches on the spectrogram—that is always, or only, associated with a single phoneme. The very same burst of acoustic energy can, for example, be heard as /sh/ when followed by an /a/ and as /s/ when followed by /u/ (Mann and Repp 1980). A /d/ is heard when there is a burst of acoustic energy of rising frequency followed by an /i/, but if the following vowel is a /u/ then a /d/ will be heard when the frequency *falls*. None of the simple features of the portion of acoustic activity that corresponds to a /d/, or to any other phoneme, are necessary or sufficient for a /d/, or whatever phoneme it is, to be heard in that acoustic activity. No simple statement can be given of these context effects, for they are found everywhere. As Liberman and Mattingly (1985) put it: '[E]very "potential" cue—that is, each of the many acoustic events peculiar to a linguistically significant gesture—is an *actual* cue' (1985: 11).

This lack of invariance is rather poor evidence for anything special about speech. Even if speech were processed in an entirely non-special way, one would not expect there to be an invariant relationship between those properties of speech sounds revealed by the speech spectrograph and the phonemes heard, for we do not, in general, expect perceptual categories to map onto simple features of stimuli in a one-to-one fashion. There is, for example, no pattern of wavelengths of light that is always or only associated with the perception of a given shade. Various wavelengths are perceived as the same shade (for there are color metamers), and the same wavelength can be experienced as a different shade in a different context (for there are color-constancy effects). Since

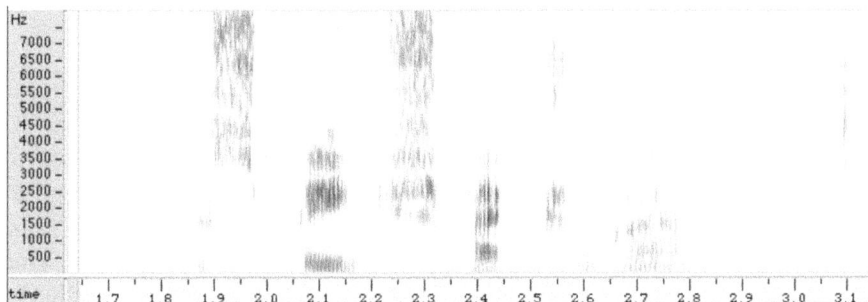

Figure 10.1. *A speech spectrogram*

Notes: *The dark patches indicate bands of frequency at which there is acoustic energy. The spectrogram shown represents an utterance of the phrase 'speech spectrogram'.*

the existence of a mapping from simple features of the stimuli to perceived categories is not the norm, the absence of such a mapping cannot be evidence of specialness. It does not illuminate the specialness of speech.

It is tempting to think that one could create the appearance of an invariance problem for *any* categorization task if one started with a sufficiently low-level description of the input. One can recognize a large number of faces viewed at various angles and in various lights, and one can recognize them beneath a wide range of hats, spectacles, false noses, and so on. The invariances which one exploits in face recognition are at such a high level of description that if one were trying to work out how it was done given a moment-by-moment mathematical description of the retinal array, it might well appear impossible. There is surely no simple pattern of retinal stimulation that always and only occurs when I see a face as being my brother's. My ability to recognize him is not a matter of my being triggered by some simple property of the array he projects to my retina. A well-trained boy scout can recognize granny knots, reef knots, and sheet bends by touch alone, but the features that he uses when making these discriminations would be extremely hard to recover from a moment-by-moment presentation of the pressure that each knot exerts on his fingertips. There is certainly no profile of finger-tip pressures that is always or only associated with granny knots, and so the same sort of invariance shown by the acoustic signal for speech is shown by the haptic signal for knots. But it would be absurd to draw any conclusions about the specialness of boy scout knot-perception on the basis of this invariance. It is simply that the boy scout does not categorize on the basis of simple features that can be discerned in the moment-by-moment description of fingertip pressures. The speech spectrographs for which we find invariance in the speech signal are moment-by-moment descriptions of the low-level properties presented to the ear. It is no surprise that they fail to show any features that correspond with the phonemes we hear in speech, and it is no indication of specialness.

Duplex Perception

Duplex perception is a strange phenomenon and it occurs in a strange context, making it rather hard to interpret, but in their article 'A Specialization for Speech Perception', Liberman and Mattingly (1989) rest their whole case for the existence of cognitive resources that are devoted solely to speech processing on the phenomenon of duplex perception. Duplex perception occurs when headphones are used to play a different sound to each ear. More specifically, it occurs when the sound given to the first ear is a speech sound: a syllable like /da/ or /ga/, but a speech sound that has been doctored so that the initial burst of acoustic energy is absent. The result of this doctoring is that the sound, if heard in isolation, is ambiguous between /da/ and /ga/. The sound which is presented to the other ear is just that burst of acoustic energy needed to disambiguate the doctored sound—the burst of rising frequency sound that, if added to the first sound, would make it sound like a /ga/; or the burst of falling frequency sound that would make it sound like a /da/. This second sound, if heard in isolation, sounds like a little chirp. It does not sound like speech. Here is Liberman and Mattingly's (1989) account of what it's like to hear this combination. (I've replaced their jargon with mine):

> Listeners hear two sounds, one at each ear. At the ear receiving the [second sound], they hear a non-speech chirp, just as they do when the [second sound] is presented in isolation. At the ear receiving the [first sound] they hear /da/ or /ga/. But, surprisingly, these latter percepts are not ambiguous, as they were when the [first sound] is presented in isolation; rather, they are unambiguously determined to be /da/ or /ga/ by the [fact about whether the second sound is a chirp of rising frequency or falling frequency], just as when the undivided syllable is presented in the normal way. (Liberman and Mattingly 1989: 490)

Perhaps this result is, as Liberman and Mattingly say, a surprising one, but is it evidence of specialness? One would not expect this mingling of the sounds presented at either ear if simple non-speech sounds were presented, but, as we emphasized above, simple sounds are not the relevant control group. To see if the duplex effect shows speech to be special, we should compare speech sounds to non-speech sounds of comparable complexity. When Fowler and Rosenblum (1990) did this, comparing the duplex perception of syllables with the duplex perception of wooden and metal door slams, the speech sounds behave in more or less the same way as the non-speech sounds. Duplex perception seems not to indicate specialness.

The McGurk Effect

The story so far is this. We are trying to understand how it is that speech perception differs from normal perception in such a way that speech sounds can be resolved much faster than other sounds. We have looked at three phenomena that are said to illuminate this specialness. Two of these phenomena (the lack of invariance and duplex perception) we found to tell us nothing about the specialness of speech *per se*. They did nothing more than point towards some ways in which the perception of complex, composite sounds can be expected to differ from the perception of simple sounds. The other phenomenon we have looked at is the categorical perception of speech. We found there to be good evidence that this phenomenon is related to the specialness that we are trying to understand, but we also saw some good reasons to doubt that this relationship is an especially intimate one. We turn now to the fourth of the phenomena that has been thought to cast light on speech's specialness. This is the phenomenon known as the McGurk effect. In the McGurk effect, the syllable that a speaker is heard to have said is found to be influenced by lip movements that the speaker is *seen* to produce, as well as by the acoustic information given to the hearer's ear (McGurk and MacDonald 1976). The effect occurs in the following way: A video is taken of a speaker repeating the syllable /ga/ and an auditory recording is made of the speaker repeating the syllable /ba/. When the auditory recording is heard alone, listeners accurately recognize the syllable heard as a /ba/. If they are hearing these syllables *while watching the video of appropriately timed /ga/s being mouthed*, then the listener is subject to an illusion in which the sound *heard* is reported as being /da/.

The illusory syllable splits the difference between the syllable heard and the syllable seen. /ba/, which gets presented to the ears, differs from /ga/, which gets presented to the eyes, in its place of articulation. /b/s are bilabial (which is to say that they are articulated at the lips), while /g/s are velar (which is to say that they are articulated towards the back of the throat). What listeners hear in the McGurk effect is a /d/, which is an alveolar consonant, made towards the middle.

The effect may be thought of as a somewhat surprising instance of the context effects that we discussed under the heading of 'The Invariance Problem'. The facts about which phonemes a burst of sound is heard to contain are, as we saw, influenced by a great many features of the *context* of the sound. What the McGurk effect shows is that context effects are not limited to effects of a sound's *auditory* context. The effect is an effect of *visual* context on heard sound. We were unmoved by the invariance problem because context effects are the norm for the perception of complex stimuli. The McGurk effect is more impressive because cross-modal context effects are less obviously normal.

But cross-modal context effects are not *entirely* exceptional. If the McGurk effect shows that there is something special about speech, it is not because there is anything special about the fact that speech is a stimulus that is subject to influence from concurrently presented visual stimuli. Lots of stimuli other than speech are subject to that sort of influence. The influence is most frequently discussed in connection with examples from outside the auditory domain, such as the illusion of self-motion produced by motion in the periphery of the visual field (Lee and Lishman 1975). Cross-modal effects are found in the auditory domain, too. The McGurk case is not the only case in which vision and auditory modalities combine in illusory ways, and so it does not show that such illusory combinations are special to speech. Saldana and Rosenblum (1993) have shown that judgments of whether a cello sounds like it is being plucked or bowed are subject to McGurk-like interference from visual stimuli. Sound and vision can also interact to produce *visual* illusions, not just auditory ones. The number of flashes that a subject seems to *see* can be influenced by the number of concurrent tones that he *hears* (Lewald and Guski 2003).[4] It is not special to speech that sound and vision can interact to produce hybrid perceptions influenced by both modalities, without the subject's being aware of the influence.

This is not to say that the McGurk effect shows us nothing special about speech. The McGurk effect does reveal an aspect of speech that is in need of a special explanation because the McGurk effect is of a much greater *magnitude* than analogous cross-modal context effects for non-speech sounds. Although non-speech sounds *are* influenced by vision in much the same way that speech sounds are influenced in the McGurk effect, they do not seem to be influenced to the same extent. The particular *degree* of influence from vision on what seems to the subject to be the auditory perception of speech does seem to be an effect that needs to be explained by the postulation of something special about speech processing.

This is worth emphasizing because a *quantitative* difference between speech perception and the perception of other sounds may be explained by reference to a *quantitative* kind of specialness on the part of speech. Given that sounds in general are *somewhat* susceptible to McGurk-like effects, we do not need to postulate very much specialness to explain why speech is distinguished from other sounds by the degree of its susceptibility to such effects. Perhaps the unusually high susceptibility of speech sounds to the McGurk effect is explained by the fact that the contexts in which speech sounds are heard are, to a greater extent than are the contexts of other sounds, occasions where the source of the sound is visible and where the visual information is a potential source of useful disambiguating information. The existence of other auditory-visual cross-modal illusions shows that there are mechanisms in place by which visual stimuli can influence the perception of sound. The fact that speech sounds are unlike other sounds in the degree to which it is *useful* to make fine discriminations, and the fact that speech sounds are unlike other sounds in the frequency with which visual information from the sound source is *available* for helping with such discriminations, could together explain why the mechanisms of cross-modal influence (not

---

[4] A vivid demonstration, described in Kamitani and Shimojo (2001), can be found at: <http://www.cns.atr.jp/~kmtn/audiovisualRabbit/index.html>.

special in themselves) come to be especially influential on the perception of speech.

The magnitude of the McGurk effect does reveal something special about the psychology of speech perception, but the specialness accounting for the McGurk effect might just be that the normal mechanisms of audiovisual interaction are especially active for speech on account of the uncommon availability of occasions on which they can come into play, and the uncommon utility of their doing so. Our final verdict on the explanandum of the Motor Theory of Speech Perception is this: the fact of speech's fast resolution needs to be explained, but the phenomena that have been discussed as if they were revealing of the ways in which speech is special turn out to tell us rather little. The McGurk effect does show something special about speech perception, but fails to make anything clear about what sort of explanation this specialness needs. It fails to tell us whether the perception of speech is special because it differs from normal auditory perception by degree, or by some qualitative difference. I want to turn now to the theory that these various phenomena are supposed to support, and that purports to give an explanation of them, and of the specialness with which we began.

2.

Any interpretation of a scientific theory is probably mistaken if the theory is interpreted as saying something trivial, or something very obvious. It is equally likely to be mistaken if the theory is interpreted as saying something obviously false. The most discussed theory of speech's specialness is the Motor Theory of Speech Perception. The task of saying what the contents of that theory are proves to be much harder than one might expect. This is not because the theory has not been given a canonical statement, but because the theory seems from some points of view to be saying something trivial and from other points of view to be saying something that is obviously false.

The canonical statement of the theory was given in 1985 when Liberman and Mattingly wrote 'The Motor Theory of Speech Perception Revised'. They tell us that 'The first claim of the Motor Theory, as revised, is that the objects of speech perception are the intended phonemic gestures of the speaker'. 'First and fundamentally', we are told, 'there is the claim that phonetic perception is perception of gesture' (Liberman and Mattingly 1985: 21). How are we to understand this claim? There are at least two possibilities, suggested by a familiar distinction from discussions of perceptual epistemology. In those discussions, we often encounter the distinction between two different perceptual relations distinguished in natural language by the difference between perceiving an entity and perceiving *that* something or other is the case. There are, corresponding to these two perceptual relations, at least two ways in which Liberman and Mattingly's claim about the object of speech perception could be interpreted. It could be a claim about the sort of thing that goes in the y place in true sentences of the form 'He heard y' (when the hearing in question is an instance of speech hearing). Or, alternatively, it could be a claim about the sort of thing that goes in place of the P in true sentences of the form 'He heard that P' (when the hearing in question is an instance of speech hearing). When Liberman and Mattingly talk of perceiving 'gestures', what they mean is that when we hear a /b/ the object of our perception is a bilabial plosive gesture; that when we hear a /n/ we hear an alveolar nasal gesture; and so on. What isn't clear is which of the two perceptual relations these gestures are supposed to be the objects of.

On a first reading, Liberman and Mattingly are claiming that when a listener hears speech, it is true that he hears intended phonemic gestures. If this is the correct reading of their claim, then their claim is surely true. Sentences of the form 'x heard s' are true if and only if there is something identical to s that x heard. This context for 's' is an extensional one. So, for example, it is true that Miss Scarlett heard the gunshot just if it is true that there is something that Miss Scarlett heard, and true that that thing was the gunshot. It doesn't matter whether she recognized it as a gunshot, or even if she has any concept of gunshots. If Miss Scarlett thought that she was hearing a champagne bottle being opened, but the sound was in fact that of a gun firing, then it is nonetheless true that Miss Scarlett heard the gunshot. She heard it; she was just mistaken about *what* she heard. Understood in this way—as a claim about the object of the x heard y relation—the 'first claim of the Motor Theory' is uncontroversial. A speech act *is* a sequence of intended phonemic gestures, so the truth of sentences of the form 'He heard intended phonemic gestures' is guaranteed by the existence of truths of the form 'He heard the speech act'. Perhaps, like Miss Scarlett, we do not know what it is that we are hearing. The claim that we hear phonemic gestures is compatible with the claim that we hear such gestures unbeknownst to us.

This claim is true, and obviously so, but it won't do as an interpretation of what Liberman and Mattingly intend, because it can't do the work that the Motor Theory is supposed to do. To claim that the perception of speech is the perception of gesture in this sense is not to identify a feature that makes speech special. Even if speech perception were exactly the same as normal audition, then the object of perception in this sense would still be the gesture. If we want the Motor Theory to be making a non-obvious claim, we should understand it to be making a claim about the other of the two sorts of perceptual relation: It must make a claim about the object of the relation 'x heard *that P*'. More is required for the truth of sentences with the form 'x heard *that P*', than was required for the truth of 'x heard s' because this context is an intensional one. Although Miss Scarlett can truly say, on learning about the circumstances of the death, 'I heard the gunshot', she cannot truly say that she heard that there was a gunshot. Not if she took it for the opening of a bottle. Hearing *that* there was a gunshot requires (among other things) that the event be heard *as being* a gunshot. If the Motor Theory claims that intended phonemic gestures are the objects of speech perception in the sense that speech perception involves perceiving *that* there were certain intended phonemic gestures, then the theory is committed to our hearing speech *as being* a set of intended phonemic gestures.

The claim that we hear speech as being phonemic gestures is rather counter-intuitive, and it is easy to produce an argument showing it to be false. Suppose we have a listener, who, being in the grip of some false theory about the phonemic gestures, believes that /b/ is not a bilabial plosive, but a postalveolar trill. Such a listener is easy to imagine. It is equally easy to imagine that such a listener is listening to a speech replete with instances of /b/, and that he is taking the experience at face value. He need not believe he is subject to any sort of illusion. If hearing the /b/s in the speech involved hearing them *as* bilabial plosives, then this thinker would be guilty of some sort of *irrationality*, just as one who believed that no gun had been fired would be guilty of irrationality if he persisted in his belief of having experienced a whole sequence of events as being gunshots. A false theory about phonetics is not so readily refuted: speech perception doesn't present us with the underlying gestures as a part of the content of experience. We could make the same point in more Wittgensteinian tones: One who is searching for a labiodental fricative may need a look-up chart to tell him when he has successfully found one. One who is searching for a red flower famously needs no such look-up chart. If phonemic gestures were given in the contents of experience, then 'labiodental fricative' would behave like 'red' in this respect. The contents of experience have to be non-inferentially *given*, and phonemic gestures aren't given in that way. It is not the case that when x perceives speech, x perceives that certain phonemic gestures were made.

It can seem that the Motor Theory is stuck with an irresolvable dilemma. Either it is making a claim about the relation of hearing, or it is making a claim about the relation of hearing *that*. If it is making the first claim, then it is saying something true, but something that cannot contribute to our understanding of the specialness of speech. If it is making the second claim, then it is saying something demonstrably false. This dilemma only arises because we take the Motor Theory to be making a claim about the object of a perceptual relation *in which the subject is a person*. Can the theory avoid these problems if it retreats to making a claim about a *subpersonal* relation? The first horn of the dilemma remains—'*x* heard *s*' is an extensional context for *s*, whatever we put in the *x* place, so the identity of speech with phonemic gesturing guarantees trivially that phonemic gestures are perceived when speech is. The move to a subpersonal perceiving subject can't help here. But perhaps it helps with the dilemma's other horn. The second horn of the dilemma does look like a place in which the tactic of moving to a subpersonal perceiving relation seems more promising. The problems at that horn were problems that arose because the theory seemed wrongly to convict a certain kind of thinker of irrationality. These are problems that the move to the subpersonal may help with, since the notions of rationality and irrationality are notions that lose their grip when we move to the subpersonal.

To avoid the problems set out above, the Motor Theory needs to be interpreted as making a claim about a subpersonal perceiving relation, and it needs this perceiving relation not to be an extensional one, or else the problems associated with the first horn of the above dilemma will arise again. How should we understand this notion of a subpersonal, non-extensional 'perceiving that' relation? When we were at the personal level, we had some intuitive grasp of the way in which the personal 'perceiving that' relation fails to be extensional, but at the subpersonal level, much more work is needed if we are to understand the source of the non-extensionality of the 'perceiving that' relation. At the personal level, hearing *that* there was a gunshot requires hearing the sound *as* a gunshot. If we are to make sense of the Motor Theory as claiming that speech is represented as phonetic gestures at the subpersonal level, then we shall need a subpersonal notion of representing *as*, corresponding to the personal-level notion of hearing *as*. It is a natural thought that the way to understand this notion is through some kind of connection with particular *concepts*. But, for reasons akin to those we've already seen, a conceptually demanding notion won't serve the motor theorist's purposes. One who lacks the concepts of phonemic gestures can nonetheless hear what's being said to him, and even if the thinker has those concepts, they do not seem to be engaged just because speech is being perceived.

The situation we are in is this: The Motor Theory makes the claim that gestures are the objects of speech perception. We are trying to understand what this could mean. We have seen that this can't be understood as being the claim that gestures are the objects of any *personal-level* perceptual relation, or of a *conceptually demanding* perceptual relation, or of an *extensional* perceptual relation. The suggestion might be made that a perceptual relation that avoids each of these problems can be built from the notion of carrying information. A subpersonal representation can *carry information about* some properties of a thing without the thinker needing concepts of that thing, and this notion of carrying information about some aspect of a thing allows us to individuate the contents of representations more finely than extension—and so it appears to enable us to find something non-vacuous to make of the idea that the representation of speech is the representation of vocalic gestures. But this appearance is misleading, and the problem of vacuity arises again. It arises because phonemes are *individuated by* the lip movements that produce them. A glance at the international phonetic alphabet will reveal that phonemes are classified by place of articulation (where in the mouth the sound is made) and by the sort of movement made (plosives, nasals, trills, taps, fricatives, and so on, are ways of moving the mouth parts). What it *is* for a word to contain a given consonant is for its pronunciation to involve mouth movements of a certain sort. Nothing can carry information about phonemes without carrying information about phonemic gestures. The phonemes that make up speech have to be encoded if one is to know what the speaker said, and so any representation that carries information about the words that a speaker said *ipso facto* carries information about the lip movements made.

The thing that is distinctive about the approach of Liberman and Mattingly and other Motor Theorists is that their notion of 'representing speech as phonemic gestures' is not tied to the notion of carrying information *about* such gestures, nor to the thinker's capacity to *think* about phonemic gestures, but instead to the thinker's capacity to *produce* such gestures in his own speech. This, finally, is the motor-related aspect from which the Motor Theory gets its name. It seems, on the face of it, that any move that links our ability to perceive speech to our ability to speak is an unappealing move, since it ought to be possible to hear speech without being able to speak oneself, and it is surely possible to hear speech sounds that one cannot produce oneself. The child born with an ill-formed mouth does not, of course, face deafness. Nor is the poor mimic unable to hear the speech of those with regional accents that it is beyond him to imitate. Liberman and Mattingly (1985) were moved by these sorts of considerations, and cite as an influence on their revision of the Motor Theory the finding by MacNeilage, Rootes, and Chase (1967) that 'people who have been pathologically incapable from birth of controlling their articulators are nonetheless able to perceive speech' (Liberman and Mattingly

1985: 24). On account of these findings, they moved from a claim about the vocal tract *itself* to a claim about an internal *model* of the vocal tract. The theory as revised does not claim that we actually use our mouths and throats in hearing speech, but that the perception of speech involves the use of 'an internal, innately specified vocal-tract synthesizer' (Liberman and Mattingly 1985: 26).

This move from a claim about the vocal tract to a claim about an internal model of the vocal tract brings with it a loss of clarity because it is not immediately obvious what it *takes* for a bit of neural apparatus to constitute an internal model of the vocal tract. Several suggestions could be made to help us understand the claim. One such suggestion would start with the observation that there are some contexts in which one system can be said to model another just if the model can be used to generate reliable predictions about the system modeled. This is the sense of 'model' in use when a load-bearing spring is said to model an inter-molecular force, the effects of LSD are said to model schizophrenia, and, perhaps, some computer programs are said to model the weather. If the Motor Theorist's claim that the apparatus of speech perception includes a model of the vocal tract is understood as a claim that involves this sense of modeling-as-prediction-generation, then problems arise along just the lines that we have already seen. There is a problem with saying that some part of our brain generates reliable predictions about vocal tract gestures if these predictions are personal-level states—normal perceivers of speech make no such predictions. And there is a problem if the 'predictions' in question are subpersonal representations encoding information about the vocal tract—any subpersonal state that encodes information about phonemes also encodes information about vocalic gestures, on account of phonemes being individuated by the vocalic gestures that produce them.

There is, however, another sense of 'model' on which the Motor Theorist's claims stand more chance of being both plausible and explanatory. We can say that one system models another if the first behaves in a way analogous to the behavior of the second, and if it does so *for analogous reasons*.[5] If one system is a model of another in this sense then it can be said to *represent* that system, and the particular states in the model that occupy the same functional role as a particular part of the system modeled can be said to represent those particular parts. This may give us a sense of 'represent' that we can use to understand the Motor Theorist's claim that we represent speech as phonemic gesture.

---

[5] This is really just a dynamic version of the common or garden concept of a model as we find it applied to model trains, and the like. It is nothing to do with the technical, logician's sense.

If one system can be said to model another in this sense (as opposed to the less demanding and already rejected sense of modeling-as-prediction-generating), then there must be a high degree of symmetry between the causal architecture of the model and that of the system modeled. If the system modeled includes two states, both of which originate from some single feature of the system modeled, then the corresponding states of the model must also share their origins. Similarly, if two states of the system modeled have *different* explanations, there should be a difference in the way the analogous states arise in the model. Moreover, where this causal isomorphism condition requires that a *single* state must feature in the explanation of some two states of affairs, that state must be a *genuinely unified* state. If the rain in Bristol and the flooding in Wales are both caused by the area of low pressure coming from the north, then my meteorological model is not good enough if its representing Bristol as rainy is a result of having access to information about rainfall, and its representing Wales as flooded is a result of having access to some quite separate body of information about river flow. The causal isomorphism requirement can't be met by gerrymandering a disjunctive state comprising both bodies of information.[6]

For the brain to contain a model of the vocal tract (and so, in this sense, for it to be able to represent speech as phonetic gestures), the system by which the brain gets from the sounds at the ear to the representation of words spoken must include a part in which the processing of representation is causally isomorphic with the treatment received by sounds as they pass from vocal chords to lips, and out. Could the brain's processing of speech proceed in a way that would satisfy this non-gerrymandered causal isomorphism requirement so that, in virtue of modeling the vocal tract, the brain could rightly be said to represent vocal tract gestures? I think not, but I do think that we have finally arrived at the correct way to understand the content of the Motor Theory of Speech Perception. The Motor Theory should be understood as a theory about the existence in the brain of a causally isomorphic model of the vocal tract. It may be that there is such a model, but, for a couple of reasons that we shall now turn to, it does not seem to be at all likely that there is.

There are two ways for the brain's processing of speech to model the vocal tract, and each is problematic. The model could work backwards—taking as input the acoustic profiles which the vocal tracts of our interlocutors put out and analyzing them to find the phonetic intentions that set the vocal tract going. This is the most obvious way for such a model to work, but the model

---

[6] Spelling out when exactly a state is genuinely unified and when gerrymandered is, of course, not an easy matter. For our purposes the intuitive notion will have to suffice.

could, alternatively, work *forwards*. The model could try out a whole range of various inputs, and use these to generate representations of various acoustic profiles which it then compares to the acoustic profile that has been encoded by the ear. When it finds a match between one of the generated acoustic profiles and the profile perceived, it can identify the input that produced the match. These two alternative ways of using a model correspond to the two strategies that, in the psychological literature, are given the unlovely names 'analysis by analysis' and 'analysis by synthesis'. An analogy will help to clarify the difference between the two approaches. Suppose that Mr Jones is playing notes, one at a time, on the piano, and that Mr Smith has the job of finding out which notes Mr Jones is playing. To help him in his task, Smith is seated in the same room as Jones, and at the keyboard of an exactly similar piano. There are two tactics Smith can use. The speediest tactic would be to lift the lid of his piano, press the sustain pedal so that the strings are not dampened, and watch to see which string resonates. This will be the string that corresponds to the note Jones is playing. The second tactic is for Smith to press each of the notes on his keyboard, one after the other, and listen to hear when the note he plays sounds the same as the note Jones plays. In each case, Smith uses his piano as a model of Jones's. The first tactic is analogous to analysis by analysis. The second tactic is analogous to analysis by synthesis. The method of analysis by analysis is the more efficient of the two.

To meet the causal isomorphism requirement, a speech processor which could successfully detect the /d/ at the beginning of 'di' and the /d/ at the beginning of 'du' would have to do so by the same means, for both /d/s result from the same pattern of gestures. But, as we saw in our discussion of the lack of invariance, the difference in the following vowel causes this pattern of gestures to produce different effects on the features of the acoustic profile. The degree to which there is a lack-of-invariance problem, as discussed above, shows that there can be no model of the vocal tract that satisfies the causal isomorphism requirement and conducts successful analysis by analysis. The causal isomorphism requirement calls for a single part of the model detecting all and only, for example, tongue-backing, while the lack-of-invariance problem tells us that there is no feature of the acoustic profile such that a device that operated as a detector of that feature would be responding to all and only tongue-backing.

Perhaps because they are aware of the tension between their claims about lack of invariance and the possibility of analysis by analysis, the advocates of the Motor Theory have typically accepted the prima-facie less plausible *analysis by synthesis* account, according to which the model of the vocal tract in the brain generates several representations of acoustic profiles, and then compares the profiles it has generated to the acoustic profile heard, so that, on finding a match, it is able to identify whatever input to the model of the vocal tract produced a representation that corresponds to the profile presented. Even if an initial bit of analysis by analysis is used to reduce the set of profiles that must be generated to a set of plausible candidates, the task of analysis by synthesis seems so vast that it could only be successfully completed in a realistic time frame if the candidate profiles are produced by massively parallel processing. There are a huge number of possible things that you could be doing with your mouth at any time, and the analysis by synthesis approach requires that a model of the vocal tract try each one of them out to see whether the acoustic consequences it generates match the sound heard. A single model of the vocal tract trying out each of these possibilities in series would have to be working at a colossal rate for speech to be perceived in real time. Analysis by synthesis is only plausible if parallel processing is employed, but parallel processors fail to meet the 'no gerrymandering' clause in the causal isomorphism requirement on modeling. To see that they must do so, suppose there are two models working in parallel, one of which tries out the lip movements corresponding to 'du' and the other of which tries out the lip movements corresponding to 'da'. On one occasion the sound presented is a 'du' and the profile produced by the first model gets matched to the profile of the sound heard. On another occasion the sound heard is a 'da' and the second model produces the match. In both cases a /d/ is recognized, and so to meet the causal isomorphism requirement there must be a single state featuring in the recognition of both sounds—but for there to be such a state is for there *not* to be separate paths operating in parallel. Analysis by synthesis is implausible unless the synthesizing models operate in parallel, but models operating in parallel fail to meet the causal isomorphism requirement, and the states of models operating in parallel therefore fail to count as representations of parts of the vocal tract.

We have tried various ways to interpret the claims of the Motor Theory of Speech Perception, but found none of them to be both plausible and meaningful as an account of how speech perception is done. We have also found that the evidence that has been thought to recommend the Motor Theory's approach is wanting. This might lead us to give the whole thing up. Nonetheless, I claimed above that I would end with a gesture in the direction of a place where these problems could be solved. That place is, I think, closer to the spirit of the original Motor Theory than it is to the more sophisticated theory that was developed in the light of the evidence and arguments that have been reviewed here. We rejected the idea that the apparatus of speech perception is the apparatus of speech production because the perception of speech that one cannot produce is so obviously possible.

We saw that this consideration led the Motor Theorists to change their claims to claims about internal *models* of the vocal tract. They went from a claim about there being just one system to a claim about two systems, one of which was a model of the other. This seems to me to have been a source of unnecessary difficulties. There was no need for the original one-system suggestion to be abandoned so entirely. The Motor Theorist can perfectly well claim that there is a single common mechanism of speech production and perception, and that this mechanism represents phonemic gestures, without being committed to the problematic idea that the capacity to produce speech always accompanies the capacity to perceive it. A single common mechanism of production and perception need only have the function of directing speech production and perception when other things are equal; it need not be the case that *whenever* the system is able to perceive it is able to produce. There are plenty of ways in which the performance of a combined production/comprehension system could be impaired on the production side alone. The Motor Theory could then claim that there are resources of speech production, that these resources represent the phonemic gestures, and that these same resources are involved in speech perception. Overlaps in processing resources for comprehension and production are a familiar and obvious idea—the lexicon, presumably, serves both, as do some resources of grammatical analysis. We can understand the Motor Theory as proposing that the overlaps in representational resources continue out to the less abstract levels of representation needed to get the mouth to move in the right way, and needed to get us into a position to know which words are said to us. This is a sketch for the sort of proposal that might be made. If it is to be developed, then we shall need to be a lot clearer about the truth conditions of the various sorts of representation postulating claims that can be made in subpersonal cognitive psychology.

References

Best, C. and McRoberts, G. (2003). 'Infant Perception of Non-native Consonant Contrasts that Adults Assimilate in Different Ways'. *Language and Speech*, 46: 183–216.

Espy, K. A., Molfese, D. L., Molfese, V. J., and Modglin, A. (2004). 'Development of Auditory Event-Related Potentials in Young Children and Relations to Word-Level Reading Abilities at Age 8 Years'. *Annals of Dyslexia*, 54: 9–38.

Fadiga, L., Craighero, L., Buccino, G., and Rizzolatti, G. (2002). 'Speech Listening Specifically Modulates the Excitability of Tongue Muscles: A TMS Study'. *European Journal of Neuroscience*, 15: 399–402.

Fowler, C. A. (1986). 'An Event Approach to the Study of Speech Perception from a Direct Realist Perspective'. *Journal of Phonetics*, 14: 3–28.

—— and Rosenblum, L. D. (1990). 'Duplex Perception: A Comparison of Monosyllables and Slamming Doors'. *Journal of Experimental Psychology: Human Perception and Performance*, 16: 742–54.

Harnad, S. (1987). *Categorical Perception: The Groundwork of Cognition*. Cambridge: Cambridge University Press.

Ivry, R. B. and Justus, T. C. (2001). 'A Neural Instantiation of the Motor Theory of Speech Perception'. *Trends in Neuroscience*, 24: 513–15.

Kamitani, Y. and Shimojo, S. (2001). 'Sound-Induced Visual "Rabbit".' *Journal of Vision*, 1(3): 478.

Klatt, D. H. (1989). 'Review of Selected Models of Speech Perception', in W. Marslen-Wilson (ed.), *Lexical Representation and Process*. Cambridge, Mass.: MIT Press.

Kluender, K. R., Diehl, R. L., and Killen, P. R. (1987). 'Japanese Quail Can Learn Phonetic Categories'. *Science*, 237: 1195–7.

Kuhl, P. K. and Miller J. D. (1978). 'Speech Perception by the Chinchilla: Identification Functions for Synthetic VOT Stimuli'. *Journal of the Acoustical Society of America*, 63: 905–17.

Lane, H. (1965). 'The Motor Theory of Speech Perception: A Critical Review'. *Psychological Review*, 72: 275–309.

Lee, D. N. and Lishman, J. R. (1975). 'Visual Proprioceptive Control of Stance'. *Journal of Human Movement Studies*, 1: 87–95.

Lewald, J. and Guski, R. (2003). 'Cross-Modal Perceptual Integration of Spatially and Temporally Disparate Auditory and Visual Stimuli'. *Cognitive Brain Research*, 16: 468–78.

Liberman, A. (1990). 'Afterthoughts on Modularity and the Motor Theory', in I. G. Mattingly and M. Studdert-Kennedy (eds.), *Modularity and the Motor Theory of Speech Perception*. Hillsdale, NJ: Lawrence Erlbaum Associates.

—— and Mattingly, I. G. (1985). 'The Motor Theory of Speech Perception Revised'. *Cognition*, 21: 1–36.

—— —— (1989). 'A Specialization for Speech Perception'. *Science*, new series, 243(4890): 489–94.

McGurk, H. and MacDonald, J. (1976). 'Hearing Lips and Seeing Voices'. *Nature*, 264: 746–8.

MacNeilage, P. F., Rootes, T. P., and Chase, R. A. (1967). 'Speech Production and Perception in a Patient with Severe Impairment of Somethetic Perception and Motor Control'. *Journal of Speech and Hearing Research*, 10: 449–67.

Mann, V. A. and Repp, B. H. (1980). 'Influence of Vocalic Context on Perception of the [sh]–[s] Distinction'. *Perception and Psychophysics*, 28: 213–28.

Miyawaki, K., Strange, W., Verbrugge, R., Liberman, A. M., Jenkins, J. J., and Fujimura, O. (1975). 'An Effect of Linguistic Experience: The Discrimination of /r/ and /l/ by Native Speakers of Japanese and English'. *Perception and Psychophysics*, 18: 331–40.

Saldana, H. M. and Rosenblum, L. D. (1993). 'Visual Influences on Auditory Pluck and Bow Judgments'. *Perception and Psychophysics*, 54(3): 406–16.

Sinnot, J. M. and Mosteller K. W. (2001). 'A Comparative Assessment of Speech Sound Discrimination in the Mongolian Gerbil'. *Journal of the Acoustical Society of America*, 110(4): 1729–32.

11

Philosophical Messages in the Medium of Spoken Language[1]

ROBERT E. REMEZ AND J. D. TROUT

1. Introduction

The psychology of spoken language offers philosophical lessons about the potential for seduction by phenomenology, and for our readiness to adopt a reductionist metaphysics in building theory. To expose these themes, we will examine the most vexing issues in the science of speech perception, drawing evidence from an assortment of ordinary cases, from speech perception by the deaf and the cochlear implant user, and from extraordinary speech perception evoked by synthetic acoustic patterns created specifically to be impossible to vocalize. We then review the way in which dominant methods in the philosophy of mind use introspection, intuition, and reflection on phenomenal experience when drawing conclusions about the nature of perceptual experience. This contemporary research represents a substantial theory about how introspection and intuition work, a theory that can be wrong and shown to be so by appeal to empirical evidence.

Speech perception research is dominated by a theoretical perspective that, witting or not, relies on a reductionist perspective that offers comfort to the intuition that speech is, fundamentally, an articulatory act of sound production, despite the linguistic governance of the creation and understanding of every utterance. According to this dominant view, fundamental linguistic units are thought to have acoustic or articulatory characteristics essential to their identity. This version of psychological essentialism, irresistible to many, reflects reductionist commitments about the real causes of language performance: that the causes are sensory or motor properties concurrent with speech. Appealing

---

[1] The authors would like to thank Michael Bishop, an anonymous referee, and the editors for useful comments on an earlier draft.

to anti-reductionist accounts in different fields of naturalistic metaphysics, we will argue that the evidence heavily favors an alternative view in which significant speech units represent homeostatic properties.

2. The Character of Utterances

The technical study of spoken language begins with a puzzle. Talkers who are mutually intelligible—if not mutually comprehensible—share a common stock of linguistic forms. They share words, speaking Anglocentrically, and it would not incite controversy to propose that they also share the syntactic aggregation of words in superordinate phrases and clauses, as well as the syllables and phoneme segments subordinate to words. Speech is intelligible when a listener resolves linguistic form, a series of segments, syllables, words, phrases, and clauses that compose an utterance. The puzzle that launches the inquiry is the absence of close correspondence between the linguistic constituents that talkers share and the properties of physical expression. The articulation that produces an utterance, its acoustic effects borne on the air that envelops talker and listener, and the auditory qualities that a perceiver experiences are unique in each instance. Because each realization is different from any other, the relation between a linguistic form and its physical expression is one-to-many. There is neither an articulatory maneuver, nor an acoustic pattern, nor an auditory quality at the core of any linguistic form (Liberman *et al.* 1967).

The research that defined the puzzle has a long history, and the search to identify criterial physical attributes of the symbolic properties of speech has often been spurred by practical as well as scientific purpose. The hierarchical nature of linguistic structure set the aims of this research. With the exception of homophones, words are distinguished by elementary units of contrast, the inventory of which composes the phoneme set of a language. English uses about three dozen of these phoneme segments expressing a dozen and a half contrast features. Typically, research on speech production and perception has taken this least constituent as its object, and in the lore that has developed among researchers, the perception of speech has come to denote the perception of the finite and small set of meaningless phoneme segments that compose the larger meaningful constituents. This has seemed like a sensible choice, inasmuch as the set of words known to a college sophomore can exceed 120,000, every one of which can be described as a series of phoneme segments drawn from the small set of three dozen (Miller 1951). Psychologically, the hierarchical nesting of structure fostered an account in which recognition of a series of individual elementary contrasts expressed in an utterance permits a listener to distinguish words from each other. Of course, a listener might just guess that a talker said SHEEP, not JEEP or CHEAP,[2] but by attending to the fine grain of the incident sound, the expressed contrast is resolved perceptually, without speculating. But, if a perceiver is able to apprehend the linguistic form of an utterance perceptually, relying neither on guesswork nor telepathy, how is this accomplished?

2.1 Reduction to Motor Types

Accounts within psychology have favored reduction of phonemes to articulatory motor categories or to auditory sensory categories. Antique descriptions stood squarely on articulatory linguistic intuition—that is, on the impressions of a thoughtful talker considering the placement and motion of articulators when producing ordinary utterances. The apparent commutability of discrete segments—that RIGHT, TRY, and TIRE are different orders of the same phonemes, spelling notwithstanding—led to a conceptualization of articulation in similar form (Kühnert and Nolan 1999). A small segmental phoneme inventory was paired with a small inventory of hypothetical vocal acts, with different concatenation orders of the articulatory ingredients responsible for articulation of different words with the same segments. If this seems like the infinite use of finite means, it is not accidental, although this characterization of articulation proved to be false. Once evidence was pursued instrumentally, the puzzle was defined.

The advent of spectrum analyzers made the fine physical acoustic properties of speech observable, and nothing like commutable segments has ever been seen. In a classic metaphor, it is as if:

> Easter eggs [are] carried along a moving belt; the eggs are of various sizes, and variously colored, but not boiled. At a certain point, the belt carries the row of eggs between two rollers of a wringer, which quite effectively smash them and rub them more or less into each other. The flow of eggs before the wringer represents the series of impulses from the phoneme source; the mess that emerges from the wringer represents the output of the speech transmitter. At a subsequent point, we have an inspector whose task is to examine the passing mess and decide, on the basis of the broken and unbroken yolks, the variously spread-out albumen, and the variously colored bits of shell, the nature of the flow of eggs which previously arrived at the wringer. Notice that he does not have to try to put the eggs together again—a manifest physical impossibility—but only to identify. (Hockett 1955: 210)

---

[2] The notation convention used here puts words mentioned as words in small caps. Phoneme segments are idealizations, indicated by slashes; phonetic transcriptions or plausible notations tied to specific instances are phonetic objects, and are indicated by brackets.

Indeed, the discrete, commutable phonetic intentions of a talker are restructured for expression as a pattern of continuous articulation. An anticipated segment or two is produced with the present one (Bell-Berti and Harris 1979), while the residue of the articulation of a prior segment or two still influences the postures and motions of the vocal organs (Liberman *et al.* 1967). In this circumstance, it is not surprising that the acoustic correlates of any individual segment vary with its proximate phonemic environment. The coincident production of conceptually successive segments is *coarticulation*, and it is one source driving the variation in correspondence between a linguistic segment and its physical expression. Although intuition conforms to the linguistic properties, namely, of discrete and serially ordered commutable segments, the articulation contradicts this simple portrait of expression. The articulation of a phoneme series is an intricate rather than a straightforward expression of the linguistic sequence, and techniques to observe articulators directly, via x-ray movies and cinefluoroscopy, brought anatomical precision to the description inferred from acoustics (Honda 1996).

Coarticulation of conceptually discrete and commutable consonants and vowels amalgamates the canonical articulatory forms of speech, conferring a graded and seamlessly progressing form upon the motor acts of expression. This merges the production of linguistically distinct consonants and vowels, obliterating the conceptual linearity of the series, and thereby blocking any model of speech production similar to typing, in which each phoneme takes an invariant articulatory form in each instance. Instead, the articulation of a given segment incorporates aspects of the preceding and succeeding units in the series, amalgamating features of its phonemic context with its own distinctive attributes. In fact, expressive variation also incorporates changes in the rate of speech, in the precision of speech, in the casualness or formality of the diction, in the intrinsic anatomical and functional differences between talkers, and in the paralinguistic expression of affect that converges in the vocal tract with linguistic expression. If the expression of any linguistically governed phoneme, for instance, /d/, varies in physical form without end, there can be no simple perceptual standard, neither articulatory nor auditory, that is adequate.

The perceptual trick required of a competent listener, according to the Motor Theory of Speech Perception (Liberman and Mattingly 1985; see also Chapter 10), is to apply intimate and expert knowledge of speech production during the perceptual analysis of the incident acoustic stream to achieve the inverse of the functions projecting phonemes into articulation and articulation into acoustics. Each listener is also a talker in this conceptualization, able to ascribe aspects of concurrent articulatory states to different moments in a speech signal and to ignore the articulatory blending of phonemes in an utterance.

The technical ingenuity of the account is admirable despite the presupposition that the phoneme, a linguistic object, is a motoric object, discoverable by unmixing the physical ingredients that the talker put into the articulatory soup.

To imagine the action of the Motor Theory, consider a listener registering the sound of an utterance and sorting the waxing and waning coarticulated influences into a tidy sequence of ordered phonemes. Because the equation of phonemes and motoric types assumed that graded subphonemic variation can be attributed solely to coarticulation, it warranted a corollary claim that subphonemic variation is imperceptible. The reason is that the motoric knowledge acting as the engine of perception as well as production completely resolves a coarticulated utterance into its initiating phoneme series (Liberman 1970). Under many test conditions, psychoacoustic research on the perceptual differentiation of synthetically produced speech revealed that subphonemic variation is barely perceptible, as this reduction of perception to articulatory types claims (Liberman et al. 1957). This research defined the study of categorical perception. In other conditions, though, subphonemic variants are well resolved despite phonemic equivalence, as in the prominent contrast between casual and careful speech, or the contrast between the speech of Flatbush and the San Fernando Valley. The existence of salient subphonemic aspects of expression undermines the asserted equivalence of phonemic and articulatory types, and reflects a convergence of influences, of which the phoneme series is one, on the properties of utterances and on the perceptually resolvable properties of speech, in consequence.

Ultimately, the claims of the original Motor Theory fell due to empirical challenges to the equation of phoneme segments and articulatory segments. Chiefly, these empirical proofs showed that speech perception can occur in conditions that neutralize the contribution of production, as in cases of degenerate, immature, or absent productive capability. In one empirical falsification, infants too young to articulate speech were reported to perceive phoneme types (Eimas et al. 1971). In another, individuals who lacked the personal experience of articulation due to neural or functional incapacity remained capable perceivers of speech (Lenneberg 1967; MacNeilage et al. 1967). By resorting to direct means of examining the intrinsic physiology of speech production, it proved to be unimaginably difficult to identify traces of hypothetically discrete, commutable motoric ingredients in a coarticulated mixture, whether the examination focused on articulatory motion, anatomical configuration, or the motor-neural signals (MacNeilage 1970).

More recently, an original approach to motoric reduction has renewed the claim of equivalence of linguistic and articulatory types. Here, a theoretical breakthrough aimed to resolve the difference in structure between the abstract linguistic segments—consonants and vowels—and the continuous, graded, and nonlinear articulation, recasting the linguistic description in a nonsegmental form (Goldstein and Fowler 2003). In this conceptualization, the phonemic properties that are expressed and recognized in speech are no more sequentially segmental than the components of articulation, relieving from the outset a prominent aspect of the discrepancy between symbolic and physical aspects of speech. In *articulatory phonology*, phoneme contrasts are conceptualized as potential shapes, placements, and phased actions of individual articulators. A word is represented phonemically in a gestural score describing the canonical form of production. By conceiving the phonology in discrete gestures that coincide temporally, the representation of linguistic properties falls into potential alignment with articulation. Lexical contrasts that had customarily received a segmental description in the grain of phonemes have been readily described as contrasts in an inventory of gestures. Under this description, a perceiver hears through the auditory effects of articulation to the graded and continuously progressing gestures that produce the sound as they mark the lexical constituents without additional elaboration or analysis. Once the shapes, placements, and phasing of the gestures are resolved, the perceiver achieves the linguistic grain requisite to identify words, at least insofar as canonical form of expression is concerned. Admittedly, an increase in realism is accomplished by designating a robust linguistic representation that inherently accommodates variation in articulatory rate as compression or expansion in the imbrication of gestures. Moreover, an emphasis on a linguistic description far closer to articulation than the traditional abstract phoneme potentially offers a natural means of explaining alternations, neutralization, assimilation, and other classic phonological phenomena. Instead of changing the symbolic composition of an expression, canonical gestural form persists, while gradient implementation of the score produces the variants.

As in the case of its predecessor, the Motor Theory, the empirical challenges to articulatory phonology are likely to determine the durability and range of its appeal. Of course, the evidence that had counted against the Motor Theory gets no traction against articulatory phonology simply because the more recent form of articulatory reduction claims no role for a faculty of production in perception. Instead, it uses productive types linguistically, to designate the phonemic form of words, and aims to relieve the mismatch in kind between the phonemic form regulated grammatically and the articulatory form responsible for expression. Nevertheless, the account seems vulnerable to challenges to the parity it asserts between canonical form and expressed form. For example, the new account predicts that the elevation of the tip of the tongue to touch the alveolar ridge of the palate behind the upper teeth at the close of the words LINT and BUTT persists covertly when LINT BRUSH is pronounced LIMP BRUSH, or BUTT KICKER as BUCK KICKER. In these cases of regressive place assimilation, typical of many English speakers, old-fashioned segmental phonology considers the departure from canonical phonemic form to reflect the underspecification of the articulatory place of the class of coronal segments. In articulatory phonology, the canonical form of BUTT KICKER retains its composition, though, with a change in phase relative to the components of the constellation, the gesture of alveolar touch occurs during the palatal closure for /k/, starkly reducing its acoustic effects to the degree that the place of the prior consonant appears to have been assimilated to the latter. In the case of LINT and BRUSH, a phase shift of gestural components permits a labial closure to conceal the alveolar touch. Empirical studies have not yet decided whether the gesture predicted in the account occurs covertly, as if the linguistic object were indeed an articulatory gesture, or whether the gesture that appears to be assimilated is absent in fact, as if the change occurs at the level of linguistic form.

During the wait for the evidence to appear, it is useful to consider that articulation is seldom a result of canonical lexical form alone. Talkers can regulate the properties of expression secondary to the choice of words and the phonemes they comprise, as in instances of vocal mimicry or disguise, or in a modulation of style (Pardo and Remez 2006). When this involves substitution of a gesture, when ROPING and RIDING become ROPIN' and RIDIN', there is more than a simple shift in the relative phase of articulators at play. In a recent example of vocal play, a child was observed expressing a portion of the range available for expressive variation without altering the lexical message. In a stable pragmatic condition, plausibly commonplace, she taunted her father: [dæri] ... [diæri] ... [dædi] ... [dæːri] ... [dæʔ di] ... [dæɾi] ... etc. Extrapolating from the language of articulatory phonology, we might say that the phonetic form related to the canonical phonemic form was edited on the run for expressive aim, although neither the lexical item nor the canonical phonemic form changed at all. The articulation followed the editing, and this unexceptional use is permitted by the principle that linguistic form and articulated form are related as the message is to its expression. They differ, and the symbolic representation of the expression cannot be reduced to its physical properties.

2.2 *Reduction to Acoustic and Auditory Types*

In a dialectic at large among researchers, a reduction of linguistic properties to motor components is typically weighed against reduction to auditory types, on the truism that speech must be heard to be understood. The germ of inspiration for sensory reduction of symbolic linguistic objects was sown a century ago in the work of introspective psychologists attempting to catalog the elementary sensations in which perceptual experience is constituted. Variation in physical frequency, amplitude, and spectrum were correlated with the elementary sensory experience of pitch, loudness, and timbre, but additional properties, among them *vocality*, were identified (Köhler 1910). Even a pure tone evoked a vocal impression, according to this theme, which hindsight might view as unduly fanciful. However, were vocality to be primary, no account would need to derive audible phonetic qualities from simpler auditory sensations.

These days, the scene has shifted, and an explanation of speech perception appealing to auditory qualities uses the warhorses of psychological rationale, similarity, and likelihood, to make the case (Diehl *et al*. 2004; Mirman *et al*. 2006; Saffran 2003). The argument, put simply, is that individual phoneme segments are differentially correlated with specific acoustic properties. For a perceiver to be sensitive to the occurrence of a phoneme, a sensory function, albeit a tacit one, must tally the distribution of the precise sensory forms affiliated now and then with each of the phoneme types. The sensory forms associated with any phoneme are asserted to vary in likelihood as well as in auditory characteristics, with a gradient of similarity extending from the likeliest sensory form to less likely forms in this view of the orderly reduction of linguistic types to auditory norms. To recognize a phoneme, its auditory form—whether simple or complex—must be analyzed and held for comparison with the different remembered sensory gradients that compose a perceiver's experience of speech. Through an act of comparison of an unidentified auditory form with the remembered gradient of auditory correlates, a perceiver can estimate the likelihood that the present sensory experience is /p/ or /t/ or /k/, or none of these. The use of an actuarial method of recognition is often disguised in the garb of neural networks. although neither neural phenotypes nor actual exposure to spoken language are modeled in such enterprises.

Could a phoneme, a linguistic marker used to distinguish one word from another, be a sensory form? Could the perception of phonemes rest on normative characterizations of sounds? These questions about speech have been posed continually since the instrumental analysis of acoustic waves was possible, and the perspective is chronic if utterly implausible, despite a lack of encouragement of auditory sensory reduction from (1) physical acoustic analyses, (2) auditory physiological studies, (3) engineering projects, and (4) psycholinguistic research. Its survival expresses a forlorn wish despite a preponderance of disconfirming evidence.

The hypothesis is false. If the phonemes of English were physical acoustic properties sensed auditorily, and were this aspect of experience informed by nothing more elaborate than a thorough actuarial practice that plotted distributions of differing likelihood, then typing on keyboards would have already been rendered obsolete by clever audio engineers and deft statisticians, in favor of vocal dictation. Sadly, these words are inscribed by a typing hand. The evidence was clear long ago that the acoustic properties of intelligible speech are not ordered normatively. Instead, it seems as if a logical function applies to sensory types including those well beyond the norm or even the physiologically possible, permitting the recognition of phonemes from auditory states that violate norms.

The claimed auditory definition of phonemes runs afoul of the hard evidence in several ways, whether the auditory component of perception is cast as a conduit faithful to the impinging spectra, or as a creative function in which criterial auditory sensory qualities are induced predictably from incipient speech. The most fundamental problem with this view is its presumption that auditory sensations are long-lasting, that is, persistent enough to permit comparison to stored distributions; and, both accurate and persistent enough to compose stored distributions.

In fact, very little of the auditory sensory effect of speech is left 100 ms after a wavefront strikes an eardrum, and nothing is left in 400 ms (Howell and Darwin 1977; Pisoni 1973). Human auditory experience is neither durably nor accurately remembered. These conclusions stem from studies of sensory acuity in which the ability to distinguish subtly different *discriminanda* was found to be inversely proportional to the interval occurring between them. In a test of this type, a listener is asked to distinguish the presentation of two physically identical sounds from two sounds of identical type differing in auditory form. When the interval between the two sounds is brief, discrimination performance can be based on a comparison of the auditory quality of the items as well as their perceived classification. As the interval grows in duration, the rapidly fading sensory trace of each item becomes less well resolved, and, therefore, less available for comparison; discrimination performance grows correspondingly poorer because at long lags it is based largely on the classification of the items, which is all that survives the lag. It is commonplace to encounter items that are discriminable at a brief lag of 50 ms yet not at a lag of 200 ms. In less than a quarter of a second, the sensory qualities are gone, leaving only the remembered experience to inform the act of discrimination. This distinction between fleeting sensory qualities and far more durable classified memory applies to sounds other than speech. Indeed, auditory quality is readily encoded as a mechanical description approximate to the physical cause of a sound (Hirsh 1988; Lakatos *et al.* 1997). Although it is possible to conjure a sound by thinking of it while the world is silent, it is implausible to suppose that this experience occurs by recalling an exact trace of a waveform once heard. More likely, the experience is generated by way of a cognitive approximation to the mechanical causes of sound.

With respect to the components of an utterance, auditory persistence is apparently worse for consonants than for vowels; however, overall, this aspect of the early sensory experience of speech is just too ephemeral to be useful for more than initiating perception. Urgency imposed by the rapidly fading auditory sensory trace of speech warrants a prompt projection into a relatively more stable phonetic or phonemic code.

Auditory accounts of phonemes have presumed that the acoustic constituents of speech spectra and the elementary auditory experiences concurrent with speech perception are finite in variety, and that the acoustic and auditory elements that promote the perception of speech form a class. If the acoustic ingredients of speech were unique to speech, this premise might be true. Yet, the whistles, clicks, hisses, buzzes, and hums that are taken to be speech attain that status by virtue of the configuration that they compose, and not because of their characteristics considered in isolation (Remez 2005). Produced alone or extracted from a speech stream, an acoustic element of speech evokes an auditory quality without an accompanying phonetic impression. In fact, the aspect of the acoustic stream that appears critical for eliciting phonetic perception is its time-varying pattern, and not the elements that compose the pattern. A perceiver even tolerates the presence of impossible acoustic elements if the spectral configuration is speechlike, and such findings ultimately falsify the claim that perception is achieved by isolated acoustic or auditory elements because of their relation to the norms of experience. Perceptual functions apparently track the causal conditions of the incident sensory forms, and can be indifferent to the sensory details that auditory reduction places at the center of the account.

Several lines of evidence converge here. In one, speech perception can be evoked by a pattern composed of a small number of pure tones. None of these can be produced by a human talker, but the pattern imposed on the frequency and amplitude variation of the tones is derived from a speech spectrum over time, and perceivers report the linguistic properties of the speech that served as the model for the tone complex; the auditory qualities are described as unnatural for a voice (Remez *et al.* 1994). In another technique, a speech spectrum is analyzed into bands roughly 1 kHz wide, and all of the detail within the band is merged. A single integrated amplitude value is derived each moment in each band, and this is used to set the power of a noise source matched in center frequency and bandwidth to the analyzing band. The result is a signal composed of a small number of wide-band noise sources changing amplitude according to the speech signal on which it is based, but without any of the acoustic elements of the original. No human vocal tract is capable of producing such a pattern, and this noise-band derivative of speech differs spectrotemporally from whispered speech in coarse and fine grain alike. This signal is also intelligible, despite an impression of a raspy voice (Shannon *et al.* 1995). In a chimerical variant of the noise-band method, the analysis is similar, though the source can be chosen freely rather than being noise. If the source is a recording of a construction site, the elements composing the resulting pattern are produced by a saw, a cement mixer, bulldozers, backhoes, shovels, loaders, fork lifts, and cranes. If the source is a jazz band, then the elements composing the pattern are trumpets, trombones, saxophones, piano, bass, and drums. The impression reported by the listeners is of the sentence from which the analysis is made, spoken by the sources, despite the impossibility of this event (Smith *et al.* 2002). Neither construction sites nor jazz bands talk, yet this is a listener's impression.

The claim that speech perception depends on resolving the normal and familiar auditory forms of phonemes is also problematic considered most broadly. In this case, the proof is provided by the speech perception prowess of the deaf. For many deaf individuals, speech reading, the visual perception of linguistic form in the absence of hearing, is quite successful, despite an utter lack of auditory sensory properties to drive perception (Bernstein *et al.* 2000). Some deaf perceivers whose speech reading is far short of perfect succeed with residual hearing despite sensitivity limited to low frequencies. Alone, such limited auditory perception is inadequate for perceiving speech, though in conjunction with vision a rather thin supply of auditory samples of speech is useful. Perhaps most remarkable is the use of an electrocochlear prosthesis to evoke auditory experience. An adult whose competence in language antedates the onset of deafness is often able to rely on auditory experience evoked in this unusual way; young children deafened before they attain linguistic ability have been observed to develop competence in perception and in production of language with this atypically caused auditory sensation (Svirsky *et al.* 2000). Once in place, an electrode causes effects that differ from ordinary auditory function, of course, but not only in the means by which neural activity is excited. For one, the inevitable spread of current along the cochlea creates a kind of frequency blur; a frequency difference as great as the interval of a minor third (the difference between the first two ascending notes of 'Greensleeves', or descending, of the 'Star Spangled Banner') is not resolvable (McDermott 2004). For another, the electrode does not penetrate to the apex of the cochlea, and is never proximate to the basilar region of lowest frequency sensitivity. This causes an upward shift of the experienced frequencies, perceptually.

Implant users who were adventitiously deafened report an impression of munchkin speech, likening the initial experience of quality elicited with the electrode to the high-pitched reedy voices of the little people in the movie, *The Wizard of Oz* (Shannon, personal correspondence). Over time, listeners say that they adapt to the odd auditory qualities of speech conveyed via an implanted electrode, though we can be confident that very little of the melodious lilt of speech is available to perception; those auditory properties are stripped from experience by the coarseness of the electrical transduction of the signal.

A tolerance of acoustic elements that transcend normal experience is actually reasonable, and must not be taken to indicate an inherent perversity in the means by which perceptual knowledge of language occurs. In the nursery and on the boulevard, the variation in physical manifestations of spoken language is vast. Anatomically, talkers range from large adult males to small children, and the variation in scale is accompanied by concomitant variation in the acoustic correlates of production. Variation in dentition and soft-tissues that affect sound production is also commonly encountered in ordinary conditions, as are the occasional individuals suffering laryngitis or rhinitis. Speech can be shouted or growled, spoken with a Partagas Robusto clenched between the teeth or spoken with edentulous gums. The distorting effects of telephones and walkie-talkies are also familiar, and speech perception survives these conditions that prohibit the veridical transmission of the acoustic products of natural vocalization. The origin of this forgiving standard of articulatory, acoustic, or auditory constituents of speech is not understood, although evidence of its action is well documented. The qualitative experience of speech depends critically on the exact nature of the acoustic elements in the signal, but the linguistic form is apprehended by following the changing pattern composed by the elements, and not the elements themselves.

The freedom from fixity of the phonetic and acoustic details that express the phonemes is inherent in the nature of the contrast on which linguistic markers depend (Pardo and Remez 2006). The key is the use of distinctive oppositions. Talker and listener alike share the small set of phonemes, and though these constituents index words, neither production nor perception must be faithful in producing the segments in their canonical form. A talker must merely indicate at any juncture which of the possible phoneme contrasts is intended, with the understanding that the listener tracking the production has the same sense of possibility in mind, a consequence of experience with spoken language. This freedom to indicate rather than to replicate the articulatory choreography specific to a word licenses a talker's use of a wide expressive range; it also permits a listener to find the auditory and phonetic attributes that compose the linguistic form of speech without a commitment to a single set of physical markers, regardless of their dimension.

2.3 Prospects of Reduction

Could reduction of linguistic phonemes to motoric or auditory elements succeed if researchers identified physical correlates of these linguistic markers with greater success? One way to hazard an answer is to acknowledge that the linguistic functions of phonemes are real, and that the prospect of reduction accordingly poor. It might be possible to correlate phoneme incidence with motor, auditory, neural, and even genetically controlled cellular phenotypes without reducing the linguistic function to its correlates. Practically, the only serious prospect available for eliminating the typist's keyboard in favor of a computer that takes dictation is to understand the conditions that allow a talker to depart from canonical articulatory form and its acoustic effects, for instance, and the dimensions of permissible departure. Likewise, there is empirical need to characterize the conditions in which a spectrotemporal aberration in a sound stream can be attributed to the talker's attempt to retain a piece of food in the mouth while articulating a poignant message. But in no case can the linguistic function of the phoneme be subsumed in the functions of articulation and auditory resolution, functions so often targeted for reductionist description appealing to the physiology of the vocal tract and of the cortex, respectively. Rather, the opposite is true—the linguistic functions of phoneme contrasts subsume their expression.

3. The Seduction of Psychology by Phenomenology and the Impression of Transparency

Phenomenological analysis begins innocently enough. We take the enormity of visual experience and examine its discernible components. The visual experience of the Taj Mahal presents it as having a color, a lightness, a shape, a texture, a location, and a duration. The auditory experience of Fido's bark has loudness, pitch (or at least a dominant one), timbre (a distinctive quality), heading, range, and duration.

No method is without presumption, and methodologies used in the philosophy of perception are not innocent. After what seems a brief flirtation with naturalism in theories of mental content, contemporary philosophy of mind is once again firmly fixed on folk explorations in the phenomenology of perception.

Although contemporary philosophy of perception makes room for many different conclusions about the nature and limits of perception, its method of inquiry, in broad outline, continues the dominant legacy in the history of philosophy: We uncover the nature of perception first by describing perceptual experience, and we gain access to that perceptual experience by introspection. We peer into or focus attention on inner experience. So the procedure is utterly intuitive. We generate a first-person report of sensory experience and then introspect its phenomenal character. Contemporary treatises in the philosophy of mind offer a digest of how this process works:

> When we introspect our experiences and feelings, we become aware of what it is like for us to undergo them. But we are not directly aware of those experiences and feelings; nor are we directly aware of any of their qualities. The qualities to which we have direct access are the external ones, the qualities that, if they are qualities of anything, are qualities of external things. By being aware of these qualities, we are aware of phenomenal character.   (Tye 2000: 51)

But the philosophical work in this vein is not merely descriptive, it is also didactic; without acknowledging the possibility of intersubjective disagreement in what we find when we look inward, it supplies the description of the experience of human perceivers in the process of sensing and introspecting:

> If you are attending to how things *look* to you, as opposed to how they are independent of how they look, you are bringing to bear *your* faculty of introspection. But in doing so, you are not aware of any inner object or thing. The only objects of which you are aware are the external ones making up the scene before your eyes. Nor, to repeat, are you directly aware of any qualities of your experience. Your experience is thus transparent to you. But when you introspect, you are certainly aware of the phenomenal character of your visual experience. On the basis of introspection, you know what it is like for you visually on the given occasion. Via introspection, you are directly aware of a range of qualities that you experience as being qualities of surfaces at varying distances away and orientations *and thereby* you are aware of the phenomenal character of your experience.   (Tye 2000: 47; emphasis in original)

> Patently, awareness of phenomenal character is not a quasi-scanning process. Our attention goes *outside* in the visual case, for example, not to the experience *inside* our heads. We attend to one thing—the external surfaces and qualities—and yet *thereby* we are aware of something else, the 'feel' of our experience. Awareness of that 'feel' is not direct awareness *of* a quality of the experience. It is awareness that is based upon direct awareness of external qualities without any inference or reasoning being involved.   (Tye 2000: 51–2; emphasis in original)

Or, to take another example, this time of an inference to a metaphysical conclusion from phenomenological considerations:

> Dispositions lie strictly outside of what is immediately perceptually presented, but colors figure in the very pith of perceptual presentation. To use a much-maligned term, colors are *given*, while dispositions are *posited*. It follows that, if colors were really dispositions, they would not be visible in the way they are. To be sure, there are dispositions associated with colors, and perhaps even inferable from them, but these dispositions are not given *in* the color. (McGinn 1996: 540)

Ordinarily, a perceptual function is nondemonstratively inferential if the ultimate structural description of the object is underdetermined by the sensory evidence (Fodor 1983). In the passage above, it is presumed that what is *given* and what is *inferred* would find consensus in any group whose members honestly reflected on the matter. Interestingly, Berkeley (1709) expressed the same confidence in *A New Theory of Vision*. And he used this premise to argue, quite falsely, that the outputs of the visual process were not arrived at by nondemonstrative inference.

There is the hope, and expectation, that an elaborate, and perhaps even technical, body of consequences can be extracted from the modest observation that phenomenal character conveys what it is like to have an experience. That expectation is implicit in the approach so widely taken in the study of phenomenal character:

> Visual experiences have phenomenal character, or more simply a phenomenology. The phenomenal character of a visual experience is what it is like to have that visual experience. In general, I will say that events of sensing, such as seeing, have a sensory phenomenology. (Siegel 2006a: 484)

The scientific study of sensation and perception may have some relevance to philosophical inquiry, but this inquiry quickly and easily erects barriers to scientific improvement and criticism of philosophical analysis: 'If the argument here is sound, our perceptual systems may include modular "input systems" of the sort described by Fodor (1983), but these systems will not be ones with which visual *phenomenology* is exclusively associated' (Siegel 2006a: 501). Our perceptual systems may be modular, but the phenomenal features of their output require no allegiance to the scientific doctrine of modularity, or, for that matter, any scientific doctrine whatever. Indeed, it may be difficult to tell, from such passages, the place of the philosophy of perception within intellectual inquiry about the mind. More important, it is difficult to tell whether the methods and findings of science have any role at all to play in addressing evidently empirical claims that philosophers make about the phenomenal character of their experience. For example, just how far-reaching is the scope of this inquiry? Can it only inform us about that part of the world that is phenomenally accessible? How does this method handle the voluminous findings that first-person reports, as well as the process of introspection, can be systematically unreliable? Are the tools of introspection so blunt that this research program, such as it is, can only tell us about the most basic distinctions in the field? Are our concepts so crude, so folk-bound, that they could not serve a taxonomic function in any project but the conceptual, first-philosophical one? And finally, what happens when intuition and scientific finding collide, as they do so often?

The phenomenological project appears to be descriptive; it attempts to detail, through introspection and report, the nature of perceptual experience, along with the mechanisms, processes, states, and properties underlying it. But it is also explanatory. It attempts to account for features of perceptual experience, once again through introspection and first-person report. Finally, the processes described in contemporary phenomenological analysis are not restricted to those with the technical vocabulary to discuss it; those processes are thought to be universal. Its characteristic treatment is that important aspects of our phenomenal experience are species-general. There is a way that phenomenology is for all perceptual experience of objects. As one philosopher puts it:

> The conclusion of this paper is that there is a phenomenological constraint on object-seeing. There is a specific sort of visual phenomenology that perceivers must have to see objects, and it is that specific sort of phenomenology that plays a role in making the situation one in which the perceiver sees a particular ordinary object o, as opposed to seeing no object at all. (Siegel 2006b: 432–3)

If commitments to transparency or intuitiveness underlie such philosophical analyses, it is difficult to get too enthusiastic about the real scope or reach of philosophy into psychology. After all, where is the evidence that putatively transparent assertions are actually accurate? Where is the evidence that the taxonomic philosophical categories of perceptual experience designated in this contemporary work are found in the assertions of psychologists of perceptual experience, or in the vernacular perceptual assertions of the citizens of the Republic? It is one thing to get swept up in a project of conceptual analysis, quite another to forget the humble roots of the inquiry, the limited tools for excavation, and the culturally local body of evidence addressed.

Psychologists have not been similarly drawn in. Indeed, psychologists put scant stock in the transparency of the *contents* of perception, and even less in the power of introspection to isolate and characterize perceptual functions. By contrast, the philosophy of perception, as currently practiced, often assumes that useful, robust, generalizable—even nomic—information can be harvested from complex sources and the percepts they prompt. This assumption proved false in physics and biology, and much of 20th century psychology has been devoted to unmasking the conceits at the foundations of judgment, problem solving, memory, and self-assessment. It is precisely these conceits on which philosophical method has relied.

In the end, it is not clear what these phenomenological analyses are supposed to show. People who use them seem to believe that they reveal more than just the trenchant intuitions cultivated by training in analytic philosophy at a particular moment in history. Yet, the standard philosophical practice of giving epistemic primacy to intuitions and their 'refinement' in reflective equilibrium has been roundly criticized in the areas of moral theory (Doris 2002) and standard analytic epistemology (Weinberg et al. 2001; Bishop and Trout 2000a, 2000b). This critical work explains how the justification of epistemic and moral claims has a structure embarrassingly similar to the basis of any ideology. As it is practiced in the English-speaking world, epistemology and moral theory reflect the favored intuitions of the West. But these intuitions and practises are not necessarily compatible with our best sciences, applicable to a wide range of human populations, or associated with documented success. If cultures of the East are less disposed to the intuition that knowledge is justified true belief, which experiments have revealed, why should we expect our analytic, Western, first-person reports and intuitions about the phenomenal character of experience to be superior? They might prove to be, but we need to see the argument and evidence. Without it, the claim to its superior reliability would be cultural prejudice of the crudest sort. Notice, for example, that the enterprise of standard analytic phenomenology includes no attempts to address similar appeals to transparency, or to phenomenality, that occur in the philosophical texts of other cultures, for example, Nyāya Sūtras.[3]

---

[3] From the Nyāya Sūtras of Gotama:

I.1.4 Perception is that knowledge which arises from the contact of a sense with its object, and which is determinate [well-defined], unnameable [not expressible in words], and non-erratic. (Radhakrishnan and Moore 1957: 359, brackets in original)

I.1.16 The mark of the mind is that there do not arise (in the self) more acts of knowledge than one at a time. (360)

Vātsyayāna's commentary (also in Radhakrishnan and Moore) on I.1.16 explains the above:

... even though at one and the same time several perceptible objects ... are in close proximity to the respective perceptive sense-organs, ... yet there is no simultaneous cognition of them; for from this we infer that there is some other cause [namely, the mind], by whose proximity cognition appears ... If the proximity of sense-organs to their objects, by themselves, independently of the contact of the mind,

This is not to say that analytic phenomenology of perception is *necessarily* compromised in the same way that much moral theory and standard analytic epistemology are. Our cognitive intuitions about our perceptual outputs may be different in just the way they need to be in order to escape this charge. But that case needs to be made.

In order to rely legitimately on first-person reports and on intuitions, we need to have evidence of their reliability beyond the fact that we can generate the reports and that we have the intuitions. They must actually be reliable, not just feel reliable. In particular, we need to have evidence that first-person reports are correct, and that they are general. With no mention of whether the same individual might respect the same constraint over time, or that the constraint is honored in the conduct of different people, we have little warrant for trusting the generalizations of phenomenal analyses. Securing this evidence would require an empirical enterprise, one that tracked the accuracy of reports and intuitions, and tested them against the reports and intuitions of individuals with very different philosophical attachments. Practitioners could respond to this rather obvious challenge by examining the strength of their methods and the scope of their conclusions. Another, complementary path would lead to experimentation on different populations. What do the people of Toraja report when asked to examine the separability of dimensions in perceptual outputs? What do Mongols say? What does Oprah say?

Two decades ago, philosophers commonly contrasted perceptual appearances with realities. Through careful experimentation, researchers uncovered underlying functions that promote the resolution of the attributes of objects and events. Some are anisotropies of sensitivity that ordinarily escape notice, though scrutiny under rarefied conditions permits more or less faithful subjective impressions. Others are less available—for instance, an experience of perceptual priming. In this experimental method, the occurrence of one event affects sensitivity to a second, independent event, as if the first event altered the threshold of detection of the second. Admittedly, this is an intuitively plausible phenomenon in olfactory sensitivity, for example, in which exposure to lemon scent immediately increases the concentration of orange scent

---

were the sole cause of cognitions, then it would be quite possible for several cognitions to appear simultaneously.   (359, brackets, ellipses in original)

And Vātsyāyana's commentary on section II.1.31 sheds further light:

When the observer cognises the tree, what he actually perceives is only its part nearest to himself; and certainly that one part is not the 'tree'. So that (when the man cognises the 'tree' as a whole) what happens is that there is an inference of it (from the perception of its one part), just like the inference of fire from the apprehension of smoke.   (366)

required for detection. Gradually, the baseline sensitivity returns. In auditory word recognition, priming is not accompanied by veridical subjective states at all. A spoken word will briefly lower the threshold for detecting other words to which it is related. Naturally, a semantic relationship is evident, in which the presentation of DOCTOR facilitates the recognition of NURSE and BREAD facilitates BUTTER; yet, neither does DOCTOR facilitate the recognition of BUTTER nor does BREAD facilitate the recognition of NURSE, indicating one requirement for priming—that is, semantic relatedness.

If this dimension of relation has the ring of plausibility, it is hardly the intuitive experience of a perceiver hearing a sentence to be flooded with impressions of the semantic relatives of the resolved words. Imagine the experience of a listener hearing mother dictate, 'Eat a little something!' decomposed perhaps as 'Absorb attack bite bolt chew devour digest dine ingest gorge nibble snack sup wolf a little tiny infinitesimal miniscule paltry insignificant something object commodity substance thing', and this example is limited, falsely, to synonymy. Additionally, a spoken word also facilitates the identification of words to which it is related only in phonemic form. The occurrence of the word SHEEP facilitates the similar words SHEET, CHIC, CHEAP, JEEP, SEEP, SEAT, SEEK, SHIP, CHIP, SHEAF, SHE, SEE, and more. None of these functions is phenomenally accessible. We can only introspect the outputs that are phenomenally accessible. The intermediate values derived in these functions do not form the ingredients of the cognitive capabilities of storage and report, producing a kind of opacity of ordinary perceptual functions and contents (Fodor 1983).

These findings have exerted a powerful influence on psychological and physiological research on the senses, but almost none on philosophical analyses of perception. Moreover, the unobservable causes of our phenomenal experience would seem explanatorily relevant, because experiences, like any complex taxonomic item, can be individuated not just by what they cause but by what causes them. Despite the obvious relevance of scientific psychology to an intellectually responsible phenomenological analysis, philosophical practitioners unburden themselves of science by insisting that they are interested specifically in the study of *experience*—the introspectible output of our perceptual systems—and so presumably not concerned with its theoretical causes or effects.

This is not to say that phenomenology cannot be part of empirical psychology,[4] but in empirical psychology the phenomenal character of experience has a far more modest and settled role. It is useful to begin with questions about

---

[4] On the contrary, good examples can be found in Noë *et al.* (2000), and Noë (2002).

how we attribute properties to objects, and how we represent those properties. But this starting place is tentative and utterly defeasible. And if we take these analyses any farther, we begin to work the philosophical theses beyond the evidence we have. We might say, for example, that our experience of color has both representational and nonrepresentational features. Our evidence might consist in examples of each, or what we would say about each case. How are we to know that a given visual experience has both representational and nonrepresentational features? Because we seem to detect this difference when we apply our first-person attention to the object.

4. Transparency

One of the most seductive features of phenomenological analyses of perceptual experience is the conviction that important aspects of perceptual output are transparent—that, no matter how short the lifetime of our sensory impressions, or how faded their traces, there is something decisive of the content that can be harvested. The perceptual material uncovered by phenomenological analysis—shape, color, timbre, bitterness, heat, alliaceousness—is patently transparent in a certain respect. When their intensity reaches human thresholds, their presence is open to casual inspection. The fact that they are present is introspectible.

Different accounts of transparency have been offered, but for our purposes the precise account does not matter. Whatever the account, the transparency of mental contents is an honored thesis in the history of philosophy. And, the integrity of phenomenological analysis presupposes *some* kind of transparency. Otherwise, there would be nothing distinctively phenomenal about the examination. Instead, the process would be a purely conceptual theoretical exercise.

Whether out of arrogance or ignorance of the empirical evidence to the contrary, a substantial class of philosophers subscribe to something like an official theory, to the effect that we have privileged access to the content of our own minds. This tradition is nicely summarized by Gilbert Ryle:

[A]ccording to the official theory, a person has direct knowledge of the best imaginable kind of the workings of his own mind. Mental states and processes are (or are normally) conscious states and processes, and the consciousness that irradiates them can engender no illusions and leaves the door open for no doubts. A person's present thinkings, feelings, and willings, his perceivings, rememberings, and imaginings are intrinsically 'phosphorescent'; their existence and their nature are inevitably betrayed to their owner.   (Ryle 1949: 154)

Descartes famously thought that he knew what he perceived and believed simply by looking inward, as it were, by immediately observing the contents of his mind. As Descartes (1637) put the point: 'nothing can be in me, that is to say, in my mind, of which I am not aware'. Some commentators on Descartes hold that he restricted his transparency thesis to *occurrent* mental states, but we shall not digress to the vicissitudes of Descartes's exegesis; our focus is on positions, not figures. In Descartes's corpus, the transparency thesis takes the form that 'there can be nothing in me, that is in my mind, of which I am not conscious' (Wilson 1981: 98).

In modern philosophy, the attachment to the transparency thesis flowed deeply beneath the surface contours that otherwise divided the landscape of early modern philosophy. Along with the rationalist Descartes, the empiricists Locke, Berkeley, and Hume all embraced versions of the transparency thesis.

Asking the reader to use introspection as a test, Locke announced that 'a man cannot conceive himself capable of a greater certainty than to know that any idea in his mind is such as he perceives it to be; and that two ideas, wherein he perceives a difference, are different and are not precisely the same' (Locke 1689). Hume, advertising a different putative benefit of transparency, consults introspection to justify the incorrigibility of transparent mental contents:

For since all actions and sensations of the mind are known to us by consciousness, they must necessarily appear in every particular what they are, and be what they appear. Everything that enters the mind, being in *reality* as the perception, tis impossible anything should to *feeling* appear different. This were to suppose that even where we are most intimately conscious, we might be mistaken.   (Hume 1739)

Bishop Berkeley's arguments for transparency may be the richest philosophically—they are certainly the timeliest—because they make the very connection between transparency and the non-inferential nature of perception that modern psychologists and some philosophers resist:

But those *lines* and *angles*, by means whereof *mathematicians* pretend to explain the perception of distance, are themselves not all perceived, nor are they, in truth, ever thought of by those unskillful in optics. I appeal to any one's experience, whether, upon sight of an *object*, he compute its distance by the bigness of the *angle* made by the meeting of the two *optic axes*? Or whether he ever think of the greater or lesser divergency of the rays, which arrive from any point to his *pupil*? Nay, whether it be not perfectly impossible for him to perceive by sense the various angles wherewith the rays, according to their greater or lesser divergence, do fall on his eye. Every one is himself the best judge of what he perceives, and what not. In vain shall all the *mathematicians* in the world tell me, that I perceive certain *lines* and *angles* which introduce into my mind the various *ideas* of *distance*; so long as I myself am conscious of no such thing. (Berkeley 1709: sect. 12; emphasis in original)

He attempts to nail down that connection by declaring, 'Since, therefore, those *angles* and *lines* are not themselves perceived by sight, it follows from Sect. X., that the mind does not by them judge of the distance of *objects*' (Berkeley 1709: 16).

The father of geometry, Euclid, had an 'emission theory' of vision: Rays of light are emitted from the eye, and objects become visible when they 'catch the rays' of the eye. In this way, vision was conceived as a species of touch. Al-Kindi, the first great philosopher of the Islamic world, resuscitated Euclid's view. He urged others, 'not to be ashamed to acknowledge truth and to assimilate it from whatever source it comes to us, even if it is brought to us by former generations and foreign peoples' (Lindberg 1971: 469). To overcome the obstacles of time and culture, the emission theory had to have considerable charm. And it did. But notice that the mechanics of visual perception were no more transparent to Al-Kindi than they were to Berkeley. If they had been, introspection alone could have revealed which of these theories of vision would have been recognized as correct.

Consider an example in the philosophical analysis of visual experience, in the service of establishing the philosophical thesis of representationalism. This thesis states that the phenomenal character of experience should be explained by its status as a species of mental representation. Philosophers have defended representationalism by appealing to the transparency of visual experience to first-person acts of attention. According to this defense, our awareness of an object passes through our experience directly to the object we are attending to. As a result, we cannot focus on the intrinsic features of our visual experience. Our attempts to do so are like trying to handle a wet fish, slipping seamlessly from the experience of blue to the object's properties of blueness or squareness.

These arguments work the apparent transparency of phenomenal processes. Some distinctly philosophical projects infer the epistemic property of incorrigibility from transparency. Other such projects infer models of psychological organization conditioned on the accuracy of our first-person access. After all, if our visual experiences are transparent, then self-observation can go proxy for experimentation. Reliance on transparency raises basic scientific and philosophical questions about the *scope* of a foundational program that attempts to analyze the constituents of perceptual outputs, the more basic elements of perceptual *experience*. If philosophical analysis proceeds with introspective tools and perceptual impressions that are culturally specific, the philosopher can do little more than armchair conceptual anthropology, cataloging and charting the perceptual experience of people similarly cultivated. This is true even when the philosopher reclining at the helm is ingenious, motivated, and equipped with specialized maps and terminology.

This view certainly brings to earth the lofty goals classically associated with certain kinds of philosophical inquiry about the mind. With its domain pared down in this way, can philosophers even achieve the goal of the psychological cataloging and charting? After all, we do seem to have access to some features of our global perceptual outputs—the appearance of a tree, or the sound of a cat's meow. Is phenomenology a fit tool for the analysis of these complex contents? In the first place, once sensory flow evokes an ultimate perceptual effect, we are left with the constructive process of *interpreting* that effect. And that has been the job of phenomenology. But interpretation of this sort is notoriously cognitively penetrable, influenced by our culturally imbued beliefs and desires. Careful philosophical analysis of scientific findings can be invaluable. But when scientific material is ignored, or replaced by intuitive material, we are left to ask whether phenomenological analysis is really just a kind of ethno-phenomenology, the anthropology of academic introspection of the West.

These cautions are timely. Because philosophical investigations have returned to phenomenological analysis, we are left to wonder whether this marks a principled reaction to the inadequacies of a naturalistic approach, or the untethered drift of interest we find in research programs in fields like literary theory. If the latter, then recent interest in old-style conceptual analysis may be largely sociological in origin. By itself, this issue is not important. It becomes significant only when philosophy begins to map the terrain of other disciplines, disguising a normative view of legitimate areas of study in a descriptive philosophical vocabulary. This occurs when philosophers extend folk phenomenology to related questions in perception. Is perceptual psychology reducible to biology? What are the constituent functions in the causal chain linking our percepts to their objects? These connections cannot be introspected. Perhaps the contents of experience, our perceptual outputs, can. But these outputs are typically multiply interpretable, and often these reports vary with culture.

In any approach that individuates properties not just by what they cause but what causes them, phenomenological analysis *just is* folk exploration into perceptual functions, not simply, if at all, perceptual experience. Disconnected from a science about the source, phenomenological analysis assumes that our experiences are classically well-defined; they assume a kind of narrow psychological essentialism.

5. The Official View: Psychological Essentialism, Old-Style

If philosophical methodology is not innocent, then different methodologies may drive us to distinct conclusions. This is a point pressed in much contemporary philosophical discussion of heuristics and biases, particularly in approaches to the problems in the philosophy of mind and biology (see especially Wimsatt 2006). For example, conventional wisdom and some enlightened theories have supposed that properties essential to species identity or, for that matter, the identity of a phoneme, have necessary and sufficient conditions, though cladistic approaches have offered an alternative account of species-membership. We may have had purely pragmatic or instrumental reasons for applying these standards. But applying them may lead more surely to reductionist viewpoints than other methods.

Phenomenological states, for example, are not individuated causally. They are individuated by their internal properties, by their sensory qualities, or by their seemings. Change the world as you like, our phenomenological states may remain unchanged. Or so we imagine (Trout 2001). Characterizing these properties as internal to the subject is hardly novel. But the observation can be worth repeating because it has been so regularly ignored. If psychological explanations were driven by seemings, those explanations would be internalist. They occur within the skin, and are presumed to supervene on their physiology.

These kinds of biases arise again in more scientific endeavors. True to our prediction, grouping or taxonomic schemes for speech sounds that are phenomenologically based lead to reductionist and old-style essentialist accounts of the objects of perception. On this view, phonetic classes necessarily or essentially possessed an acoustic invariant, and there was thought to be a one-to-one mapping of taxonomic speech units onto reduction bases like either vocal gestures, intended gestures, or acoustic classes (Appelbaum 1999, 2004). This essentialist fiction gets a foothold from the phenomenal character of a speech unit presented in isolation, but why has it flourished?

5.1 *The Doctrine*

If you consult Ladefoged's (2005) classic resource for phonetics, we find a catalog of phonetic segments. These segments can be arrived at by comparing two sequences in a minimal pair, such as MAT and MAD. This list treats each segment as though it belongs to a class that is distinguished by an acoustically invariant feature. Each phonetic segment is an island; no part of any segment is part of another. This impression is abetted by the fact that the minimal-pair difference that defines each class is phenomenologically accessible: You can hear the difference between the [t] and the [d] in MAT/MAD. But hearing the difference is not the same as identifying distinctive features, because you can do the former without doing the latter. However, this does not change appearances. From the appearance that phonetic segments are assembled serially /m/ /æ/ /t/, people infer that they are spoken serially, with clearly identified boundaries.

But speech is not produced in that way. Instead, the vocal gestures that produce these features occur simultaneously; they are coarticulated. For example, the /m/ in MAD is produced by simultaneously closing and releasing the lips with the velum lowered while the larynx is buzzing; to produce /æ/, the blade of the tongue is lowered in the front of the mouth and the root of the tongue is drawn forward in the laryngopharynx, again, while laryngeal action produces the buzz of phonation. The lowering of the jaw required for the vowel cannot occur until the labial closure and release required for the /m/ are accomplished, but the tongue shape appropriate for the vowel can be set well in advance of the labial release and the lowering of the jaw—this description of the concurrent production of subjectively sequential segments is typical of articulation. This general character of articulation is unlike the subjective impression of linear concatenation of separate segments in the language's phonology (Fant 1962; Liberman *et al.* 1967).

5.2 *Underlying Functions and Psychological Opacity*

Perceptual inference is a complex matter. In the case of sound perception, we saw that there are levels of coding involved as the sound percept evolves from the sensory periphery to neural centers, and then connects to the cortical territory that guides our motor functions. These are codes that have a chemical and biochemical vocabulary, but not one that we think in, or ones that we speak. They do not have the properties necessary for our transparent experience of them. They are not accessible to the mechanisms of storage and report. We cannot make them the objects of our attention.

Berkeley's account of transparency, then, presupposed a now discredited view of this coding process. Given the psychological opacity of most of perception—even aspects of its output—we cannot infer that a psychological function does not exist just because we are unable to detect it upon introspection or reflection. While the absence of a psychological function could explain our failure to detect it, there might be other explanations. If we were able to peer into any of our perceptual functions at any stage, at any time we liked, we might delay their appointed rounds, and so slow their operation tremendously.

But a good part of the value of nontransparency resides in the fluency of perception it captures. Perceptual functions are shielded from the meddling of attention and memory, so that they can discharge their action-guiding duties, sometimes with life-saving quickness.

The level of function, and the unavailability to awareness of its details, makes the intermediate qualities opaque to introspection.

6. Conclusion

If explanations for spoken language understanding do not proceed by appeal to well-defined phenomenal attributes or to biological features of talkers and listeners, what items are implicated? Most scientific explanations account for effects by invoking natural kinds—classes of objects that play a taxonomic, counterfactual-supporting role in explanation. Open sets of properties, processes, states, and objects can endure over time even though they depend on a changing environment. Homeostasis offers a way of explaining the stable covariation of these properties. Animals, for example, are designed to maintain certain internal states, such as a body temperature range. With these features in place, the stability may invoke an even wider circle of properties. Consider a group of properties associated with birds: a high metabolic rate, feathers, hollow bones, nest building, and flight. There are birds that lack one or another of these characteristics. But homeostatic property clusters have the effect of favoring the presence of some properties, and by doing so erect barriers to the reception or development of others (Boyd 1999).

Put generally, in a homeostatic kind, no single property is a necessary condition for membership in the kind. If the homeostatic account accurately depicts the relationship among the properties in a natural kind at least in biology, psychology, and the social sciences, then we have an explanation for the field's resistance to reduction. At a minimum, we must abandon old-style essentialism. Our foregoing discussion of phenomenal attributes in speech perception stands as a disconfirming instance of the claim that explanatory kinds are classically well defined, at least in one area of psychology, and as a confirming instance of a homeostatic view of natural kinds in that field.

If our analysis is fair, then phonemic classes are not defined by necessary and sufficient conditions—not by gestural components, auditory qualities, or phenomenal characteristics. They are defined only by their role as contrast markers distinguishing words within a language. As we argued in Section 2, while their phenomenal properties may suggest as much, an utterance is not a set of articulator shapes and motions, nor sound segments strung together like commutable beads. The conditions for understanding an utterance tolerate huge variation in the structure of a vocal tract, in the nature of articulatory action, and in concurrent nonlinguistic aim. When it comes to the expression of taxonomic kinds like phonemes, class membership is open. The ultimate linguistic meaning of a kind of spoken sequence depends on properties that are diagnostic, but not extensionally definitive, of particular speech sounds. These homeostatic properties may be deeply theoretical. And theoretical states and processes are neither transparent, nor routinely accessible, to intuition.

And if our analysis is useful, we will learn much more about the nature of the mind when we use a scientific methodology in studying durable issues about the mind. If introspection and phenomenological analyses have a place in intellectual inquiry, their accuracy must first be justified using methods that do not presuppose their reliability. The range and reliability of introspection—at least about short-lived sensory states—are amply documented in the history of psychophysics, and there the evidence supports the careful use of introspection in philosophical analyses of sensory experience. But this is a modest subset of the domains in which introspective insights have been pressed for philosophical gain. Our own philosophical tradition has held at one point or another that nearly every mental state was transparent to introspection and that our intuitions were high-fidelity guides to accurate theories about the mind. That tradition in philosophy remains a formidable obstacle to knowledge in the study of the mind.

References

Appelbaum, I. (1999). 'The Dogma of Isomorphism: A Case Study from Speech Perception'. *Philosophy of Science*, 66: S250–S259.

—— (2004). 'Physical Segments and Functional Gestures', In A. Agwuele, W. Warren, and S. Park (eds.), *Proceedings of the 2003 Texas Linguistics Society Conference*. Sommerville, Mass.: Cascadilla Proceedings Project.

Bell-Berti, F. and Harris, K. S. (1979). 'Anticipatory Coarticulation: Some Implications from a Study of Lip Rounding'. *Journal of the Acoustical Society of America*, 65: 1268–70.

Berkeley, G. (1709). *A New Theory of Vision*. New York: E. P. Dutton & Co.

Bernstein, L. E., Demorest, M. E., and Tucker, P. E. (2000). 'Speech Perception without Hearing'. *Perception & Psychophysics*, 62: 233–52.

Bishop, M. and Trout, J. D. (2005a). *Epistemology and the Psychology of Human Judgment*. New York: Oxford University Press.

——— (2005b). 'Pathologies of Standard Analytic Epistemology'. Noûs, 39: 696–714.
Boyd, R. (1999). 'Homeostasis, Species, and Higher Taxa', in R. A. Wilson (ed.), *Species: New Interdisciplinary Essays*. Cambridge, Mass.: MIT Press, 141–86.
Descartes, R. (1637). *Discourse on Method, Optics, Geometry, and Meteorology*, P. J. Olscamp (trans.). Indianapolis, Ind.: Bobbs-Merrill.
Diehl, R. L., Lotto, A. J., and Holt, L. L. (2004). 'Speech Perception'. *Annual Review of Psychology*, 55: 149–79.
Doris, J. (2002). *Lack of Character*. New York: Cambridge University Press.
Eimas, P. D., Siqueland, E. P., Jusczyk, P., and Vigorito, J. (1971). 'Speech Perception in Infants'. *Science*, 171: 303–6.
Fant, C. G. (1962). 'Descriptive Analysis of the Acoustic Aspects of Speech'. *Logos*, 5: 3–17.
Fodor, J. A. (1983). *Modularity of Mind*. Cambridge, Mass.: MIT Press.
Goldstein, L. and Fowler, C. A. (2003). 'Articulatory Phonology: A Phonology for Public Language Use', in A. S. Meyer and N. O. Schiller (eds.), *Phonetics and Phonology in Language Comprehension and Production: Differences and Similarities*. Berlin: Mouton de Gruyter, 159–207.
Hirsh, I. J. (1988). 'Auditory Perception and Speech', in R. C. Atkinson, R. J. Hermstein, G. Lindzey, and R. D. Luce (eds.), *Stevens' Handbook of Experimental Psychology, Volume I: Perception and Motivation*. New York: Wiley-Interscience, 377–408.
Hockett, C. F. (1955). *Manual of Phonology*. Bloomington, Ind.: Indiana University Publications in Anthropology and Linguistics, 11.
Honda, K. (1996). 'Organization of Tongue Articulation for Vowels'. *Journal of Phonetics*, 24: 39–52.
Howell, P. and Darwin, C. J. (1977). 'Some Properties of Auditory Memory for Rapid Formant Transitions'. *Memory & Cognition*, 5: 700–8.
Hume, D. (1739). *Treatise of Human Nature*, D. F. Norton and M. J. Norton (eds.). Oxford: Oxford University Press, 2001.
Köhler, W. (1910). 'Akustische Untersuchungen, II' [Acoustic investigations]. *Zeitschrift für Psychologie mit Zeitschrift für Angewandte Psychologie und Charakterkunde*, 58: 59–140.
Kühnert, B. and Nolan, F. (1999). 'The Origin of Coarticulation', in W. J. Hardcastle and N. Hewitt (eds.), *Coarticulation: Theory, Data and Techniques*. Cambridge: Cambridge University Press, 7–30.
Ladefoged, P. (2005). *A Course in Phonetics*, 5th edn. Belmont, Calif.: Heinle/Thomson.
Lakatos, S., McAdams, S., and Causse, R. (1997). 'The Representation of Auditory Source Characteristics: Simple Geometric Form'. *Perception & Psychophysics*, 59: 1180–90.
Lenneberg, E. H. (1967). *The Biology of Language*. New York: Wiley & Sons.
Liberman, A. M. (1970). 'Some Characteristics of Perception in the Speech Mode', in D. A. Hamburg (ed.), *Perception and its Disorders*. Baltimore: Williams & Wilkins, 238–54.

Liberman, A. M., Cooper, F. S., Shankweiler, D. P., and Studdert-Kennedy, M. (1967). 'Perception of the Speech Code'. *Psychological Review*, 74: 421–61.

——Harris, K. S., Hoffman, H. S., and Griffith, B. C. (1957). 'The Discrimination of Speech Sounds within and across Phoneme Boundaries'. *Journal of Experimental Psychology*, 54: 358–68.

——and Mattingly, I. G. (1985). 'The Motor Theory of Speech Perception Revised'. *Cognition*, 21: 1–36.

Lindberg, D. C. (1971). 'Alkindi's Critique of Euclid's Theory of Vision'. *Isis*, 63: 469–89.

Locke, J. (1689). *Essay Concerning Human Understanding*. Oxford: Oxford University Press.

McDermott, H. J. (2004). 'Music Perception with Cochlear Implants: A Review'. *Trends in Amplification*, 8: 49–82.

McGinn, C. (1996). 'Another Look at Color'. *Journal of Philosophy*, 43: 537–53.

MacNeilage, P. F. (1970). 'Motor Control of Serial Ordering of Speech'. *Psychological Review*, 77: 182–96.

——Rootes, T. P., and Chase, R. A. (1967). 'Speech Production and Perception in a Patient with Severe Impairment of Somesthetic Perception and Motor Control'. *Journal of Speech and Hearing Research*, 10: 449–67.

Miller, G. A. (1951). *Language and Communication*. New York: McGraw-Hill.

Mirman, D., McClelland, J. L., and Holt, L. L. (2006). 'An Interactive Hebbian Account of Lexically Guided Tuning of Speech Perception'. *Psychonomic Bulletin and Review*, 13: 958–65.

Noë, A. (2002). 'Experience and the Active Mind'. *Synthese*, 29: 41–60.

——Pessoa, L., and Thompson, E. (2000). 'Beyond the Grand Illusion: What Change Blindness Really Teaches Us about Vision'. *Visual Cognition*, 7: 93–106.

Pardo, J. S. and Remez, R. E. (2006). 'The Perception of Speech', in M. Traxler and M. A. Gernsbacher (eds.), *The Handbook of Psycholinguistics*, 2nd edn. New York: Elsevier.

Pisoni, D. B. (1973). 'Auditory and Phonetic Memory Codes in the Discrimination of Consonants and Vowels'. *Perception & Psychophysics*, 13: 253–60.

Radhakrishnan, S. and Moore, C. A. (1957). *A Sourcebook in Indian Philosophy*. Princeton: Princeton University Press.

Remez, R. E. (2005). 'Perceptual Organization of Speech', in D. B. Pisoni and R. E. Remez (eds.), *The Handbook of Speech Perception*. Oxford: Blackwell, 28–50.

——Rubin, P. E., Berns, S. M., Pardo, J. S., and Lang, J. M. (1994). 'On the Perceptual Organization of Speech'. *Psychological Review*, 101: 129–56.

Ryle, G. (1949). *The Concept of Mind*. Chicago, Ill.: University of Chicago Press.

Saffran, J. R. (2003). 'Statistical Language Learning: Mechanisms and Constraints'. *Current Directions in Psychological Science*, 12: 110–14.

Shannon, R. V., Zeng, F. G., Kamath, V., Wygonski, J., and Ekelid, M. (1995). 'Speech Recognition with Primarily Temporal Cues'. *Science*, 270: 303–4.

Siegel, S. (2006a). 'Which Properties are Represented in Perception?' in T. Szabó Gendler and J. Hawthorne (eds.), *Perceptual Experience*. Oxford: Oxford University Press, 481–503.

—— (2006b). 'How Does Visual Phenomenology Constrain Object-Seeing?' *Australasian Journal of Philosophy*, 84: 429–41.

Smith, Z. M., Delgutte, B., and Oxenham, A. J. (2002). 'Chimaeric Sounds Reveal Dichotomies in Auditory Perception'. *Nature*, 416: 87–90.

Svirsky, M. A., Robbins, A. M., Kirk, K. I., Pisoni, D. B., and Miyamoto, R. T. (2000). 'Language Development in Profoundly Deaf Children with Cochlear Implants'. *Psychological Science*, 11: 153–8.

Trout, J. D. (2001). 'Metaphysics, Method and the Mouth: Philosophical Lessons of Speech Perception'. *Philosophical Psychology*, 14: 261–91.

Tye, M. (2000). *Consciousness, Color, and Content*. Cambridge, Mass.: MIT Press.

Weinberg, J., Nichols, S., and Stich, S. (2001). 'Normativity and Epistemic Intuitions'. *Philosophical Topics*, 29: 429–60.

Wilson, M. (1981). *Descartes*. London: Routledge and Kegan Paul.

Wimsatt, W. (2006). 'Reductionism and its Heuristics: Making Methodological Reductionism Honest'. *Synthese*, 151: 1–31.

Index

4'33"; see also Cage, J. 142–4
a-spatial theories 9–11, 97–8, 100–101, 108–109
acousmatic experience; see also musical listening 16, 58, 66, 146–8, 150, 158–9, 164, 166–7, 170–1, 173, 175, 177, 178, 179
acousmatic thesis 15–16, 48, 50, 51, 50, 61, 63, 64, 167–70, 172, 175, 179
Adorno, T. 173
agnosia, auditory 106
agnosia, visual 106
Appelbaum, I. 257
Aristotle 27–8, 126, 132
articulatory gestures 19, 20, 206, 216, 234
articulatory categories 236–8
articulatory phonology 239–4
attention 16, 58, 66, 81, 123–5, 155, 158, 160, 162, 169, 171–2, 177, 186, 190, 203–4, 205, 206, 247, 253, 255, 258–9
attention, joint 207
audibilia 57, 60
audible qualities 18, 31–2, 35, 42, 46
auditory experience
  of space 32, 84–5, 87–9, 101–4, 107, 112, 165
  of empty places 89, 136, 174
  of sounds as located 8–11, 28–33, 39, 77, 83–4, 88, 90, 98, 101, 107, 174, 203
  contrasted with visual experience 85–8
auditory scene analysis; see also grouping 62–4, 72–5
auditory streaming 62, 105, 107, 160

Balashov, Y. 141
Bataille, G. 142
Batty, C. 2
Beckett, S. 64
Belin, P. 206
Bell-Berti, F. 237
Bellmann Thiran, A. 106, 107
Bennett, J. 37, 61
Berkeley, G. 33, 34, 104, 133, 248, 254, 255, 258
Bernstein, L. E. 244
Best, C. 216
Bigand, E. 13

Bishop, M. 234, 250
Blanchot, M. 142
Blauert, J. 10, 28, 82
Block, N. 54
Boghossian, P. A. 91
Boyd, R. 259
Breen, N. 47
Bregman, A. S. 13, 22, 57, 62, 63, 72, 74, 79
Burkert, W. 153

Cabe, P. 70
Cage, J. 142–4, 169
Capgras Syndrome 47
Carello, C. 70
Casati, R. 6, 7, 9, 10, 77, 97, 98, 100, 101, 104
categorical perception 179, 194, 213–6, 219, 238
Chion, M. 58, 65, 155, 157, 169, 178
Chomsky, N. 198
Clark, A. 129, 130, 137
Clark, P. 146, 163
Clarke, S. 81, 82, 106, 107
Coarticulation 237–8
cochlear implants 262
cocktail party effect 105
'coming from'; see also location of sounds 12, 29, 102–3, 111, 117, 125, 137, 165, 176–7, 178, 203, 205, 208
Cox, C. 146, 155, 163
Crane, T. 4
cross-modal illusions 2–3
cross-modal perception; see also multi-modal perception 178, 206, 220–1

Dack, J. 154, 157
Darwin, C. J. 81, 243
Davidson, D. 61
Davies, P. 142
Davies, S. 146, 162
De Lorenzo, R. 176
deaf hearing 81–2, 106
Dehaene-Lambertz, G. 194, 195
Dennett, D. C. 54
Descartes, R. 254
Deutsch, D. 63, 79
Dhomont, F. 154
Diehl, R. L. 241

266 INDEX

distal theories of sound; *see also* location of sounds 8–9, 10–11, 97–8, 99, 100, 109
disturbance 27, 28, 36, 39, 41–8, 50, 51, 52, 57, 69, 71, 72, 117, 118
Divenyi, P. L. 63
doppler effect 99, 165
Doris, J. 250
Dorman, M. 135
Dretske, F. 104, 127
Driver, J. 83
duplex perception 213, 218–9
duration of sounds 5, 11, 28, 30–2, 37, 45–6, 57, 59, 67, 111, 117–8, 132, 246

echoes 26, 44–8, 99, 140
  argument from 44
egocentric spatial content 81, 83, 98, 101, 102, 103, 106–9
Eilers, M. 176
Eimas, P. D. 238
electronic music 104, 147, 152, 154, 159, 163, 165, 167, 173
Emmerson, S. 154
Espy, K. A. 216
essentialism 234, 236, 257–9
Evans, G. 2, 101, 179, 199
event theories of sounds 3–4, 5–6, 13, 28, 36–9, 50–1, 77–8, 97, 98–100, 100–3, 173–4

Fadiga, L. 211
Fant, C. G. 258
Fletcher, N. H. 71
Fodor, J. A. 54, 190–1, 201, 248, 252
Fowler, C. A. 20, 212, 219, 239
Freed, D. J. 70
frequency components 71–6, 78–80, 82, 204
Fukuda, S. 204
function of auditory perception; *see* hearing, function of

Gallagher, S. 2
Gaver, W. W. 70
Gelfand, S. A. 28, 29, 57
gestalt; *see also* grouping 57, 62–3, 131
Gilchrist, A. 92
Goldstein, L. 239
Goodale, M. A. 106
Grice, H. P. 138, 139
grouping 57, 66, 72–5, 78–81, 105, 205
  role of space in 78ff.
Guski, R. 221

hallucination, auditory 22, 189
hallucination, of silence 127
Hamilton, A. 7, 8, 13, 15, 16
Handel, S. 13
harmonics; *see also* overtones 71, 73, 78, 153, 156, 162, 163, 164
Harnad, S. 215
Harris, J. 237
Harris, K. S. 237
Harrison, J. 153
Harvey, J. 146, 162, 163, 165, 169, 170, 173, 176
hearing-in 146, 150, 175
hearing, function of 13, 52, 57–8, 69–70, 175, 204, 208, 213
Heathcote, A. 146, 155, 172
Helmholtz, H. 52
Hirsh, I. J. 63, 242
Hobbes, T. 99
Hockett, C. F. 236
Holmes, T. 152
homeostatic kinds 259
homeostatic properties 20, 235, 259–60
Honda, K. 237
Howell, P. 242
Hughes, H. C. 131
Hukin, R. W. 81
Hume, D. 254
Hurvich, L. M. 126

illusion, auditory 2, 3, 7–9, 22, 29, 30–1, 36, 46, 105, 165, 178, 189, 205, 220
  motion bounce 135
  scale 63, 79
  McGurk 220, 224
information 3, 11, 13, 20, 31–2, 35, 37, 52, 57, 60, 64, 65, 99, 105–6, 108, 134, 148–9, 157, 159, 170, 176, 183, 185, 191, 194–5, 196–7, 200, 220, 221
interference 141–2, 42–4
introspection 128–31, 247
  unreliability of 11, 20–2, 249, 254–6, 259
Ivry, R. B. 216

Jack, A. 21
Justus, T. C. 216

Kahn, D. 156
Kamitani, Y. 221
Kaplan, D. 76
Katz, D. 93
Keeley, B. L. 128

Kershaw, S. 28
Kim, J. 61
Klatt, D. H. 212
Kluender, K. R. 215
Kohler, W. 241
Kostelanetz, R. 143
Kostov, V. 204
Kuhl, P. K. 215
Kuhnert, B. 236
Kunkler-Peck, A. 70

lack of invariance 213, 216–8, 219, 229
Ladefoged, F. 257
Lakatos, S. 242
Lane, H. 216
Lee, D. N. 220
Lee, R. K. 164
Lenneberg, E. H. 238
Levinson, J. 149
Lewald, J. 221
Li, X. 95
Liberman, A. M. 19, 195, 212, 217, 218–9, 222–7, 235, 237–8, 258
light 8, 35, 56, 59, 80, 92, 99, 112, 114–7, 119–21, 255
lightness 92, 246
Lindberg, D. C. 255
Lindsay, G. 197
linguistic experience 19, 189
linguistic information 17, 183, 196–8
Lippman, E. 148, 149, 159, 166, 167, 168, 174, 175, 177, 178
Lishman, J. R. 220
Located Event Theory of sounds 97, 98–100, 100–1, 105, 109
location of sounds 7–9, 10–11, 14, 21, 28–32, 37, 77–8, 82, 97–8, 102, 103, 106–8, 118, 119, 122, 126, 136–8, 175–9, 183, 202–3, 205, 208
locational hearing 31
Locke, J. 5, 27, 51, 53, 254
Lopez, F. 149
loudness 5, 11, 20, 26, 28, 30–2, 34, 35–7, 43, 44, 52, 57, 69, 71, 129, 136–7, 141, 170, 204, 241, 246
Lutoslawski, W. 179
Lycan, W. 2

MacDonald, J. 178, 196, 220
Maclachlan, D. L. C. 5
MacNeilage, P. F. 226, 238
Mallock, A. 160
Malpas, R. M. P. 9

Mann, V. A. 217
Manning, P. 156
Marslen-Wilson, W. 190
Martin, M. G. F. 2, 76, 83–4
masking; *see also* spatial deafness 81–2, 106–7
Matthen, M. 7
Mattingly, I. G. 217, 218, 219, 222–3, 226–7, 237
McAdams, S. 13
McDermott, H. J. 244
McDowell, J. 17, 186–93, 196–202, 207–9
McGinn, C. 55, 56, 248
McGurk effect; *see also* illusions 178, 196, 220, 224
McGurk, H. 178, 196, 220
McRoberts, G. 216
meaning
    as public 186–8
    hearing of 17–8, 155, 170, 183–4, 189, 197, 204–7
meaningful sounds 65, 66, 148–9, 171, 190, 235
medial theories of sound 97–100, 109
medium-dependence of sound 6, 27–8, 33–6, 39, 99, 137, 140
Miller, G. A. 235
Miller, J. D. 212, 215
Milner, A. D. 106
Mirman, D. 241
Mishkin, M. 106
missile-like sound 29, 39
Miyawaki, K. 216
Moore, C. A. 250
Morgan, R. 163
motor theory of speech perception 19, 206, 211ff., 234, 237–9
multi-modal perception; *see also* cross-modal perception 10, 16, 101, 148, 177, 178
music 2, 4, 6, 14–6, 17, 38, 50, 64, 66–7, 81, 104, 107, 127, 132, 142, 144, 146ff.
    as autonomous 154, 170, 177
musical experience 2, 6, 22, 78, 147, 149–51, 157, 174, 179
    non-auditory aspects 149, 157–8, 160–9
musical listening 4, 14–5, 66, 81, 101, 149, 154–7
    twofold thesis 169–73
musique concrete 146, 147, 151–7, 158–9, 170, 172, 173

Nettl, B. 167
Neuhoff, J. 13
Nevo, F. 139
Nolan, F. 236

noise 36, 105, 124, 129, 143, 156, 158, 161–2, 169, 170, 176, 178, 184–6, 189, 190, 191, 197, 212, 243–4
Nudds, M. 7, 9–10, 13, 26, 44–7, 103–4, 105, 136, 174, 177–8, 183, 203–4, 208
Nyman, M. 143, 144

O'Callaghan, C. 5–9, 13, 26, 36, 50, 77, 100, 126, 132, 137–8, 203
O'Shaughnessy, B. 2, 7, 9, 102–3, 114, 203
Odor 128, 133–4
Oldak, R. 70
olfaction 2, 251
overtones; *see also* harmonics 52, 57, 161–6

Palombini, C. 153–4
Pardo, J. S. 240, 245
Pascal, B. 135
Pasnau, R. 5, 7–8, 26–8, 33, 35, 51, 77, 126, 137–8
perception
  at a distance 112–4, 117–25
  directional 113–4, 176
  mediated 114, 190
  indirect 3, 14, 104, 126
phenomenology, use in philosophy of perception 20–21, 246–52
phoneme 18–19, 134, 141, 191, 194–7, 201, 206, 212–8, 220, 226–7, 235–47, 257, 260
physicalism 6, 20, 51, 58, 174
Piorkowski, R. L. 240
Pisoni, D. B. 242
pitch 5, 11, 20, 26, 28, 30–7, 45, 52, 57, 63, 67, 69, 129, 136–7, 141, 149–50, 153, 156, 159–64, 170, 176–7, 213–4, 241, 245, 246
Pittenger, J. B. 70
Price, H. H. 128
priming 251–2
property theories of sound 5–6, 13, 27, 33–6, 38, 40, 137
proximal theories of sound 8–9, 10, 97–8, 100, 103, 109
psychological essentialism 234, 256, 257–59
psychological function 258
pure event 6, 15, 50, 61–5, 67, 160, 173–4
purely auditory experience; *see also* Strawson, P. F. 2, 9, 16, 101, 136, 166, 174, 179, 180
Pylyshyn, Z. 22
Pythagoras 58, 146, 151–7, 172

Quine, W. V. 185–9

Radhakrishnan, S. 250
reductionism, in psychology 234–46
Remez, R. E. 11, 19, 20–21, 240, 243, 245
representation 2, 5, 85, 171, 190
  indeterminate 86, 88
representational theory of sound perception 124
Repp, B. H. 217
Rey, G. 19
Richards, W. 70
Robinson, J. M. 130
Roepstorff, A. 21
Rosen, C. 168
Rosen, G 26
Rosenblum, L. D. 219, 221
Rossing, T. D. 71
Russell, M. 96
Ryle, G. 253

Sadie, S. 154
Saffran, J. R. 241
Sanford, D. 133
scale illusion 63, 79
Schaeffer, P. 58, 146–7, 152–61, 166, 170, 178
Schafer, R. M. 149
Scherer, K. R. 204
Schiff, W. 70
Scholl, B. 1
Schwitzgebel, E. 11, 21
Scott, M. 2
Scruton, R. 6, 13, 15, 16, 26, 50, 54, 55, 58, 62, 64, 66, 146, 148–51, 154, 156–60, 163, 167–8, 170–4, 176
secondary object 6, 15, 50, 58–61, 65, 67, 173, 176
secondary quality 1, 3, 5, 6, 27
Sekuler, R. 135
Sethares, W. 161
Shannon, R. V. 244, 245
Shimojo, S. 221
Siegel, S. 248–9
silence 7, 38, 57, 119
  duration of 127, 132–3
  location of 136–41
  perception of 5, 126ff
  as proper object of hearing 133
  intermodal effects of 135
Smalley, D. 154, 162, 173
smell; *see also* olfaction 2–3, 5–6, 59–60, 118, 133–4, 174

Smith, A. D. 128
Smith, B. C. 7, 9, 13, 17, 18–9, 200, 207
Smith, Z. M. 244
Sorensen, R. 5, 7, 126, 134, 140
sound-generating events; *see also* sound sources 37, 40, 42
sound-object (*objet sonore*) 154, 156, 159, 165
sound
  art 149, 151, 156, 158, 160
  generation 41
  reproduction 152, 154, 178
  sources 3–4, 6–16, 18, 21, 26, 28–33, 38, 40, 62, 69–84, 89, 91, 93, 94, 97–8, 103, 107–8, 132, 137–8, 142, 153, 156, 157, 165, 170, 175, 177–8, 183, 191, 196, 203–5, 208, 243–4, 250
  stream 169, 170, 183, 193, 194, 206, 216, 243, 246
  waves 10, 51–2, 56, 60, 83, 99, 100, 105, 120, 122, 123–4, 136, 139, 141–2, 143, 198, 204, 205, 208, 241
sounds
  as events 3, 6, 13, 15, 26ff., 50ff., 77, 97–100, 103–5, 109, 160, 173
  as individuals 15, 38, 76, 165, 171
  as abstract individuals 7, 76
  as objects of hearing 4–6, 17, 22, 26, 30, 52, 57, 65, 104, 109, 148, 158, 169, 203, 222
  as particulars 26–8, 36, 38, 44, 45–8, 76, 107, 137
  as pattern types 76
  as phenomenal entities 111, 122
  as properties 3–5, 11, 13, 26–7, 33, 35–6, 39, 48, 51, 60, 91–4
  as traveling 7–8, 117–8, 120, 203
  as waves 4, 6–9, 26, 117, 119, 123–4, 137–9, 140, 176, 203
  at a distance 8–10, 28, 76, 112ff., 140
  contrast with colors 92–4
  double-duration of 117–9
  individuation 5, 20, 60
  information in 8–10, 43, 48, 71–3, 83, 204–5, 208
  instantiation 7, 75–7
  location of; *see* location of sounds
  movement of 122, 138, 150, 164–5
  not spatially individuated 80
  production of 14–6, 41, 146, 147–51, 157–8, 160, 162, 168, 172–3, 177–9
  spatial structure of 80–1, 102–3
spatial audition 78–91, 98–109
  necessarily egocentric 83

spatial deafness 81–2, 106–7
speech
  as special 14, 17, 21, 194, 204, 212–3, 215–7, 219–22, 224
  easter egg analogy 236
  perception 17–21, 190, 194, 195, 205–6, 211ff., 234ff.
  perception, objects of 209, 222–5
  understanding 183–7
speed of sound 44–6, 102, 120
Spence, C. 83, 205
Spitzer, M. 146, 148
stimulus meaning 185–6
Stockhausen, K. 152, 164–6, 173
Strawson, P. F. 2, 9, 60–1, 101–2, 103, 107, 136, 147, 174–5, 179
Svirsky, M. A. 244
synesthesia 131

Thomson, W. 162, 170
timbre 5, 11, 13, 20, 26, 28, 31–2, 34–7, 52, 56–7, 67, 69, 129, 136, 141, 156, 170, 171, 176–7, 213, 241, 246, 253
  as non-acousmatic 160–9
tone 13, 36, 63, 67, 129, 131, 137, 148, 150–1, 161–2, 171, 213, 215, 221
  pure 161, 173, 176, 241, 243
  vs. music 156
  vs. noise 170
  vs. sound 64, 158
touch 2, 10, 59, 83–5, 133, 174–5, 178–9, 218, 255
transmission 26, 39–42, 48, 60, 79, 93, 140, 154, 245
Trout, J. D. 11, 19, 20, 21, 194, 250, 257
Turvey, M. T. 70
Tye, M. 5, 247
Tyler, L. 190
Tyrrell, J. 154

Ungerleider, L. 106
Urmson, J. O. 132

vacuums 6, 39, 48, 100, 137
  argument from 33–6
Van Leeuwen, T. 149
Velleman, J. D. 91
VenDerveer, N. J. 70
Verbrugge, R. R. 70
vibration 5, 6, 27, 33, 35–6, 41–3, 52, 57, 60, 66, 67, 73, 77, 137, 141, 143, 163, 166, 174, 175, 214
  and character of sound 93, 161

vibration (cont.)
  information embodied in 70–2
  patterns of 7, 19, 75, 77, 216, 234
virtuosity 16, 166
visibilia 59, 60
visibility of light 114–7
vision 1–3, 8, 10, 12–4, 22, 57, 106, 133, 135, 178, 220–1, 244, 248, 255
  spatial character of 80, 83–4, 88
vocality 241
voice 18, 28, 65–6, 149, 154, 161–2, 163, 168, 169, 170, 183, 184, 191, 194, 195, 196, 204, 207, 208–9, 243–4, 245
  information conveyed by 204–6
voice-onset time (VOT) 213–4
von Kriegstein, K. 196, 206
Vouloumanos, A. 21

Walton, K. 167
Warner, D. 155, 163
Warnock, G. J. 126

Warren, R. M. 70, 195, 196
Watson, D. 147
wave theory of sounds; *see also* sound waves, sounds as waves 6–9, 36–36, 38–48, 75–6, 122, 137–9, 140
waves 70–1
  and earthquakes 138–9
Weinberg, J. 250
Werker, J. F. 21
Wiggins, D. 61
Wilder, A. 147
Wildes, R. 70
Wilson, M. 254
Wimsatt, W. 257
Windsor, L. 158, 170
Wishart, T. 152, 158, 165
Withington, D. 176
Wittgenstein, L. 53–4, 196, 224
Wollheim, R. 146, 150, 171
Worby, R. 174, 178

Zuckerkandl, V. 148

The manufacturer's authorised representative in the EU for product safety is
Oxford University Press España S.A. of el Parque Empresarial San Fernando de
Henares, Avenida de Castilla, 2 – 28830 Madrid (www.oup.es/en or product.
safety@oup.com). OUP España S.A. also acts as importer into Spain of products
made by the manufacturer.

www.ingramcontent.com/pod-product-compliance
Ingram Content Group UK Ltd.
Pitfield, Milton Keynes, MK11 3LW, UK
UKHW021317180426
11947UKWH00015B/1290